AI
ADOPTION

AI

ADOPTION
STRATEGIES AND
TACTICS FOR
SUCCESS

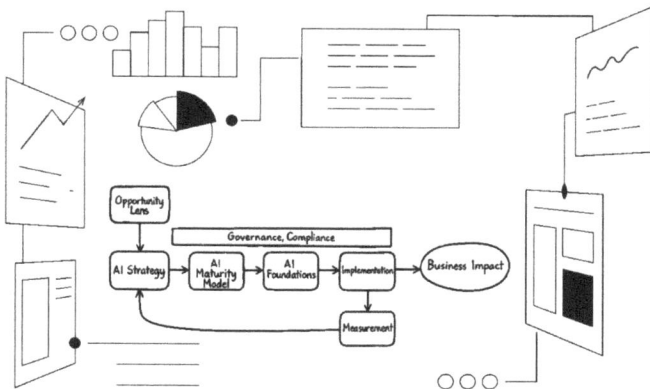

M. M. (SATH) SATHYANARAYAN

Contents

Acknowledgments

This book reflects the insights and candid guidance of leaders across the entire AI adoption spectrum—from board directors and C-suite executives to operations leaders, academics, and hands-on practitioners as well as contributions from independent research organizations and AI solution vendors that offered critical case studies and market perspectives.

Don Fowler—a seasoned board member and CEO of multiple high-tech companies—provided invaluable encouragement and strategic perspective. His input ensured this book addresses the questions boards and investors most care about when evaluating AI's long-term impact on enterprise value.

Ion Nemteanu—formerly Sr. Director of Data Science, Marketing Intelligence, and Software Engineering at Thermo Fisher Scientific, then Executive Director of Business Analytics at UCSD's Rady School of Management, and now founder of Nemsee Consulting—brought an extraordinary blend of operational leadership and academic rigor. Ion performed an end-to-end review of the manuscript, drawing on his experience directly implementing enterprise AI initiatives and shaping data-driven business transformations. His review and insights were invaluable. His feedback helped refine the concepts and go deeper in some cases, and also confirmed continuity of flow across chapters from an executive point of view.

Dr. Sridhar Mitta—founder and managing director of NextWealth and former head of Global R&D at Wipro—is the recipient of many honors and awards, most recently the Dataquest Lifetime Achievement Award. With 50 years driving technology innovation across multinationals and entrepreneurial ventures, he brought the perspective of a seasoned board-level leader. He reviewed early drafts, drawing on his experience steering product and services strategy for global clients, and his engagement added a valuable technology and business lens to this work.

Julian Gingold—a board member—read the drafts with a sharp eye for strategic oversight, helping ensure the narrative stayed aligned with the considerations leadership teams face when assessing AI's risks and rewards.

Bob Henderson—a CFO, CMC board member, and consultant with deep experience in manufacturing finance—reviewed early drafts and corroborated this book addresses the priorities financial leaders care about most. His background in FP&A, supply chain, and IT transformation brought critical perspective on linking AI initiatives to enterprise value. As a trusted advisor to executive teams, he helped ensure the guidance here remains both strategic and practically grounded.

GB Singh—Director of Business Management at Solar Turbines— reviewed the manuscript in full and offered thoughtful encouragement on the perspectives shared. With more than two decades of global leadership across energy systems, power generation, and large-scale industrial initiatives, he brought a valuable, seasoned viewpoint to this work. His engagement is genuinely appreciated.

Chirayu Vaidya—Chief Delivery Officer at NextWealth and formerly Global Vice President of Operations at Jumio—brought the perspective of an operations executive who has led thousands across global delivery centers. He reviewed the manuscript and verified it addresses the challenges and priorities faced by leaders driving large-scale service delivery and process excellence. His experience managing multi-industry operations and customer success initiatives worldwide added meaningful depth to this work.

Dinakar Nallapureddy—an experienced data science and optimization leader with a track record of improving large-scale supply chain and warehouse operations—read early drafts and validated that this book speaks to the practical realities faced by technical teams. His background managing machine learning and simulation initiatives across industries brought a valuable on-the-ground perspective. I deeply value his thoughtful engagement.

Dan Faggela—CEO of Emerj Artificial Intelligence Research—was kind enough to grant permission to draw from Emerj's extensive research library and case studies. His work provided a broad, cross-industry lens on AI best practices that helped enrich the strategic foundations of this book.

Bhavin Shah—CEO of Moveworks—offered valuable vendor perspectives and facilitated access to practical case studies from the front lines of enterprise AI adoption. His input added an important dimension on how advanced AI applications are deployed within organizations.

Jeff Richardson—an AI strategy and innovation consultant with extensive experience across healthcare, insurance, manufacturing, and technology—reviewed the manuscript and noted how closely its frameworks align with his own perspective on practical AI implementation. His background in guiding organizations through collaborative, human-centered, AI initiatives provided a valuable, cross-industry lens. I'm grateful for his thoughtful validation of the ideas presented here.

Lily Poursoltan—a data scientist at UC San Diego Health—generously shared insights from her work developing and operationalizing advanced AI models for patient flow and emergency care. Her perspective on translating data science into day-to-day decision-making offered valuable lessons on the realities of implementing AI in complex environments. I am grateful for her openness and thoughtful input.

Hyo Duk Shin—Professor of Innovation, Information Technology, and Operations at UC San Diego's Rady School of Management—reviewed early drafts and provided thoughtful encouragement on the material. As faculty director of the Institute for Supply Excellence and Innovation, he also facilitated access to case studies that enriched this work. I welcome his support and the practical perspectives he brought to this effort.

Bijan Zayer—President of Zayer & Associates and long-time professor of leadership and organizational strategy—reviewed the manuscript and was supportive of the material and approaches presented. I appreciate his encouragement and interest in this work.

Paul Barton—Americas Region Sales Manager at HP and adjunct professor at CSUSM—leads OEM partnerships and sales growth across the United States, Canada, and Mexico. I treasure his willingness to review the manuscript and the credibility his leadership perspective brings to this work.

Finally, I want to acknowledge the many executives, data scientists, and operational leaders, named and unnamed, whose frank discussions about both successes and setbacks in AI adoption have enriched this work. In addition, authoritative resources like the *Harvard Business Review* provided invaluable perspectives on organizational change, leadership, and innovation that helped frame many of the concepts discussed here. Their collective willingness to share hard-earned lessons ensured this book remains firmly grounded in the realities of strategic AI adoption.

A Note on the Writing Process

In keeping with the theme of this book, I incorporated generative AI tools as part of the writing process—not as a substitute for authorship, but as a collaborator in shaping, refining, and structuring ideas. Every insight, framework, and perspective remains my own. This work reflects what responsible AI adoption looks like: human-led, tool-assisted, and always grounded in clarity, judgment, and purpose.

On Copyright and Source Use

Every effort has been made to ensure that all references, quotes, and third-party materials used in this book are either cited appropriately or fall under permissible use. Where external sources have informed the work, they are acknowledged either within the text or in the bibliography. All diagrams and frameworks are original unless explicitly noted. Any resemblance to other materials is coincidental or derived from widely available public concepts that have been reframed through the author's own lens.

Introduction

Leading AI-Driven Business Transformation

This book is a practical guide for business leaders who want to turn artificial intelligence (AI) from a buzzword into a business driver. Whether you're exploring AI for the first time or trying to move beyond isolated experiments, this book will help you navigate adoption challenges, align AI initiatives with business goals, and deliver measurable impact. Through structured frameworks, real-world case studies, and actionable methodologies, you'll learn how to lead AI adoption with clarity, confidence, and strategic intent.

AI is no longer an emerging trend. It's already reshaping how companies work, communicate, and create value. From marketing and customer service to product development and operations, AI-driven solutions are improving key business metrics: increasing productivity, reducing costs, enhancing customer satisfaction, and driving revenue growth. Generative AI (Gen AI) has accelerated this transformation, moving from experimental tools to operational solutions in record time.

But while this first wave of AI is driving significant productivity gains, a larger shift is on the horizon: the rise of the autonomous enterprise. This next frontier goes beyond AI supporting decisions; it envisions AI systems that drive decisions, self-optimize processes, and operate with minimal human intervention. The potential is enormous, but so are the challenges.

For every successful AI initiative, there are dozens that stall in pilot phases or fail to scale. Why? Because AI introduces unique management challenges that differ from traditional technology implementations:

- Continuous oversight: AI models evolve as data changes, requiring ongoing monitoring, refinement, and governance.

- Nondeterministic behavior: Unlike rule-based software, AI generates outputs based on probabilities, not certainty.

- Unpredictability in language models: Tools like ChatGPT may produce different responses to the same input, raising consistency concerns.

- Hallucinations and misinformation: Gen AI can confidently produce inaccurate or misleading information.

- Bias in decision-making: AI systems can perpetuate or amplify data biases, posing reputational, operational, and regulatory risks.

Beyond these AI-specific issues, organizations face broader barriers: limited AI understanding among senior leaders, unclear strategies, poor data quality, talent shortages, resistance to change, and growing concerns around privacy and compliance (as highlighted by Bojinov, 2023, and Vizard, 2024).

As someone who has spent decades leading business transformations and helping organizations navigate disruptive change, I wanted to cut through the AI hype and confront the real, often messy question: How can leaders adopt AI in a way that delivers meaningful, scalable business impact—beyond pilots, amid challenges, and aligned with strategic goals?

This book is the result of that exploration. It brings together lessons from my experience leading change, advising industry leaders, and conducting targeted research into AI's unique opportunities and obstacles. Along the way, I had the opportunity to refine these ideas while teaching a course for business leaders at the University of California, San Diego (UCSD).

Rather than offering one-size-fits-all answers, this book provides structured frameworks and methodologies to help you navigate AI adoption with clarity and confidence. You'll find practical insights on:

- The benefits AI can deliver—and the challenges to expect

- Where to begin, and how to set realistic goals

- Preparing your organization for successful AI implementation

- Measuring success and sustaining momentum

AI adoption is complex, but it can be demystified with the right approach. My goal is to help you make informed, context-specific decisions that drive real business value (see Figure I.1).

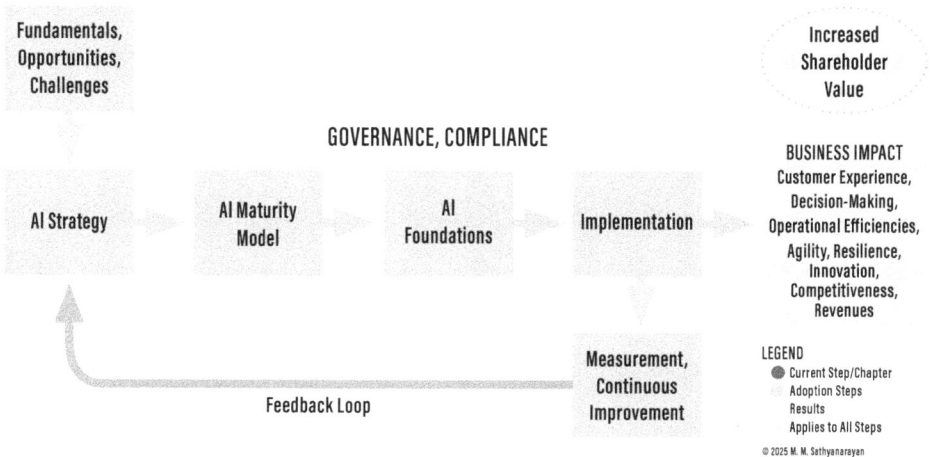

FIGURE I.1 AI ADOPTION ROADMAP

To guide you through this journey, the book is organized around a seven-stage framework that builds progressively from understanding to impact:

- Understand AI fundamentals, opportunities, and challenges (Chapters 1 and 2).

- Develop a clear AI strategy aligned with business objectives (Chapter 3).

- Assess AI readiness using the AI Maturity Framework (Chapter 4).

- Build strong foundations in infrastructure, talent, culture, and processes (Chapter 5).

- Implement AI solutions with targeted projects that demonstrate value (Chapter 6).

- Establish governance and ensure ethical, legal, and regulatory compliance (Chapter 7).

- Measure AI success and refine strategies to sustain long-term value (Chapter 8).

Each stage is explored in detail, with actionable insights and real-world examples to help you translate strategy into execution.

The journey to successful AI adoption starts with understanding both the opportunities and the obstacles ahead. Chapter 1—"AI Opportunities and Challenges"—begins by examining how organizations are using AI today, and what business leaders need to know to move from experimentation to enterprise impact.

Chapter 1

AI Opportunities and Challenges

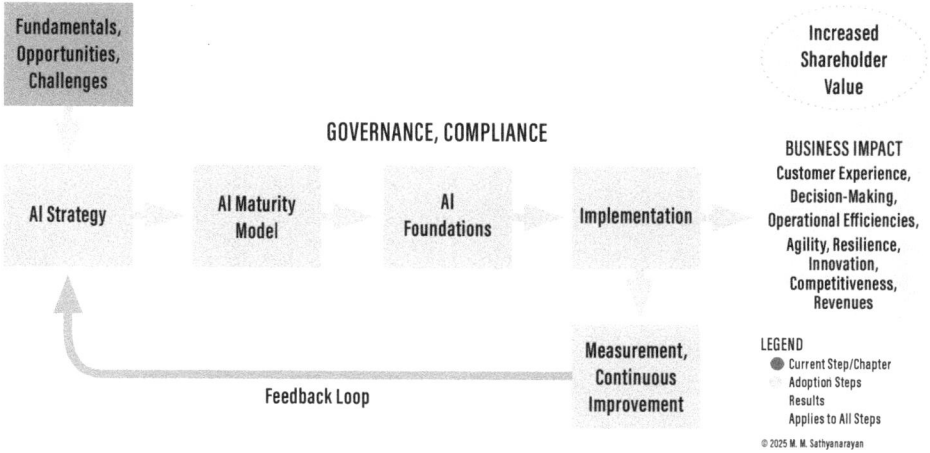

FIGURE 1.1 AI ADOPTION ROADMAP

AI is reshaping how organizations operate, compete, and innovate. But real adoption success depends not just on the technology, but on identifying the right opportunities—and anticipating the roadblocks that stand in the way (see Figure 1.1). This chapter begins that journey by examining where AI is already making an impact, and the challenges that can derail progress.

AI Opportunity Lens

From a functional perspective, AI introduces five core capabilities that open new opportunities for solving business problems and creating value:

1. Prediction of Future Outcomes
   - Use Case: Forecasting trends, risks, and opportunities based on historical data.
   - Example: A retailer uses AI-powered sales forecasts to streamline supply chains, reducing stockouts by 20%. Expanding predictive modeling applications into industries like finance (e.g., credit risk assessment) and healthcare (e.g., patient outcome predictions) further demonstrates AI's transformative potential.

2. Handling Unstructured Data
   - Use Case: Extracting insights from complex, unstructured sources like social media, videos, and text.
   - Example: A legal firm uses AI to scan and summarize thousands of documents, reducing research time by 50%. NLP models like GPT and BERT are pivotal in processing unstructured data, enabling deeper insights and automation of repetitive tasks.

3. Learning and Adaptation
   - Use Case: Adapting in real time to new data and environments.
   - Example: A cybersecurity platform refines its algorithms with each detected attack to improve threat detection. AI systems use reinforcement learning to dynamically adjust to new threats in environments such as fraud detection and network security.

4. Interaction with Humans
   - Use Case: Enhancing user experiences with intuitive interfaces.

- Example: AI chatbots handle routine inquiries, improving customer service. As Juniper Research (2023) notes, voice commerce is rising rapidly, signaling broader consumer readiness for AI-driven interaction channels.

5. Operating at Scale and in Real Time
   - Use Case: Processing massive datasets for dynamic decision-making.
   - Example: An airline optimizes ticket pricing dynamically, increasing revenue by 10%. AI-driven operations management tools enable real-time decision-making in logistics, traffic management, and financial markets.

The following examples highlight how AI is being applied across seven core business functions, illustrating not just what's possible, but how different industries are already making it real.

Marketing and Sales: *Personalization at Scale*

Core Challenge
Customers now expect personalized experiences tailored to their preferences, behaviors, and context. Traditional systems, however, are not designed to learn or adapt dynamically. They rely on static rules and fixed datasets, making it difficult—if not impossible—to deliver the kind of evolving personalization that today's consumers demand.

AI in Action
AI enables personalization at scale by continuously learning from customer behavior, predicting needs, and generating tailored content. It moves beyond static segmentation to deliver relevant

experiences across touchpoints, automatically adapting campaigns as new data becomes available. This shift allows marketing teams to move from reactive messaging to proactive customer engagement.

Key Technologies
Machine learning (ML), natural language processing (NLP), Gen AI (text, image, video), predictive analytics, computer vision, reinforcement learning

Business Impact
- Increased productivity in marketing teams
- Enhanced personalization reduces churn and boosts revenue
- AI-powered campaigns drive higher engagement and conversions

Industry Adoption
Global brands are leveraging AI-driven marketing to improve customer engagement and campaign performance:

- Coca-Cola used Gen AI in its campaigns to streamline creative development and deliver visually rich content at scale.

- Amazon's recommendation engine is a cornerstone of its retail strategy, accounting for an estimated 35% of purchases through real-time, personalized product suggestions, as noted by Jarrell (2018). Its use of dynamic AI pricing also contributes to its competitive edge, as highlighted by Symson (2023) and Prisync (2023).

- Netflix relies on AI-based personalization to reduce churn and increase viewer retention, with its recommendation system reportedly saving the company over $1 billion annually.

AI's role in marketing is now a critical enabler of personalization at scale. Organizations that adopt AI-driven approaches are seeing measurable improvements in customer engagement, campaign effectiveness, and operational efficiency. As AI capabilities mature, businesses that embed personalization into the core of their marketing strategy will be better positioned to meet evolving customer expectations.

Software Development and IT: *AI-Augmented Engineering*

Core Challenge
Software development is resource-intensive, often slowed by long backlogs, debugging cycles, and the complexity of maintaining legacy systems. As demand for new features grows, traditional methods struggle to meet speed and quality expectations, creating bottlenecks that delay time to market.

AI in Action
AI is increasingly used to augment—not replace—engineering teams. It automates repetitive, low-value tasks across the software development lifecycle, helping developers focus on higher-impact work. Strategic use of AI can accelerate delivery, improve code quality, and reduce operational overhead, though human oversight remains essential—particularly for security, architecture, and complex logic.

Use Cases Across the Lifecycle
- Requirements and Planning: Summarizes user stories and identifies specification gaps using NLP and large language models (LLMs).

- Design: Recommends design patterns based on prior projects (ML, knowledge graphs), though lacks contextual architectural judgment.

- Implementation: Coding assistants generate boilerplate code and complete functions from natural language prompts (LLMs, NLP).

- Testing: Auto-generates test cases and predicts failure points (ML, predictive analytics), improving coverage but requiring manual quality assurance for edge cases.

- Deployment and DevOps: Optimizes Continuous Integration and Continuous Delivery (CI/CD) pipelines and flags anomalies in real time (reinforcement learning, anomaly detection).

- Maintenance: Detects bugs, suggests refactoring, and supports log analysis and security monitoring.

Key Technologies
LLMs, NLP, ML, reinforcement learning, static and dynamic code analysis, anomaly detection, AI-enhanced CI/CD tools

Business Impact
- Accelerates development timelines
- Improves code quality and test coverage
- Increases developer productivity and reduces cognitive load
- Enhances operational monitoring and reliability
- Strengthens application security when properly integrated

Industry Adoption

Organizations with mature DevOps practices and cloud-native architectures are leading adoption:

- GitHub Copilot is widely used for code generation in enterprise and open-source projects.
- Google and Microsoft deploy internal AI tools for code review and bug prediction.
- A 2023 McKinsey survey found over 25% of software leaders in large enterprises actively use AI in development and testing.

Considerations for Leaders

AI adoption in software engineering is progressing but remains bounded by key limitations: lack of contextual awareness, potential inefficiencies in complex logic, and the risk of introducing insecure code patterns. Effective adoption requires balancing AI automation with strong development practices, human validation, and secure workflows. The opportunity is significant—teams that integrate AI thoughtfully are seeing gains in speed, quality, and output.

Customer Service and Support:
Intelligent Virtual Agents

Core Challenge

Customer service centers face overwhelming inquiry volumes, high support costs, and inconsistent service quality. Human agents are limited in capacity, prone to fatigue, and struggle to meet 24/7 support demands. Routine inquiries slow response times, while complex issues get backlogged, leading to dissatisfied customers.

AI in Action

AI-powered virtual agents enhance customer service by resolving routine queries and assisting human agents with real-time recommendations. AI chatbots and voice assistants handle common tasks like password resets, order tracking, and product information, allowing human agents to focus on more complex or sensitive issues. These systems also support efficiency gains through smarter call routing, sentiment detection, and proactive engagement.

While AI excels at handling predictable, repetitive tasks, it remains limited in its ability to understand emotional nuance or solve problems that require context, creativity, or negotiation. Poorly implemented bots can frustrate users—especially when escalation paths to live support are difficult to access or unclear. Successful implementations pair AI with fast and seamless hand-offs to human agents when needed, blending scale with empathy.

Key Technologies

Natural language understanding, LLMs, dialogue management systems, speech-to-text and text-to-speech, sentiment analysis, reinforcement learning

Business Impact

- Improves efficiency by offloading routine requests from human agents
- Reduces support costs by scaling self-service options
- Enables faster response times and greater service consistency
- Augments live agents with recommendations and knowledge retrieval
- Contributes to better customer experience when well integrated with human support

Industry Adoption

- Companies across sectors are deploying AI to streamline support operations and improve service quality. Bank of America's "Erica" virtual assistant now handles millions of customer requests. E-commerce brands like H&M rely on AI to manage product questions and delivery updates. Airlines are using AI to analyze calls in real time and surface relevant information to agents, improving first-contact resolution.

- What sets apart successful implementations is not the technology itself, but how it is deployed. Businesses that combine automation with thoughtful escalation, continuous learning, and clear customer pathways are seeing better outcomes in customer satisfaction and operational agility.

Finance and Risk Management:
Smarter Fraud Detection and Decisions

Core Challenge

Financial institutions face escalating challenges in fraud detection, credit assessment, and regulatory compliance. The growing volume, velocity, and complexity of financial data can strain internal teams, slow down critical decision-making, and increase exposure to operational, financial, and reputational risk.

AI in Action

AI is reshaping finance and risk management by helping institutions detect anomalies, automate decisions, and monitor compliance more proactively. ML models can analyze transaction patterns in real time to flag potential fraud while reducing false positives. In credit underwriting, AI enables more adaptive and inclusive scoring approaches—particularly useful for applicants with limited credit history or nontraditional financial data.

In capital markets, AI supports investment strategies by processing large volumes of structured and unstructured data—including market signals, news, and economic indicators. In compliance, NLP tools assist by reviewing communications, contracts, and disclosures to identify potential violations or irregularities faster than manual processes.

However, the deployment of AI in finance must be accompanied by robust explainability measures. Regulatory frameworks, such as the EU's AI Act and the U.S. ECOA, require that AI-driven decisions, especially those affecting consumers, be transparent and justifiable. Financial institutions must ensure that AI models provide clear insights into their decision-making processes to comply with these regulations and maintain stakeholder trust.

Key Technologies
ML, graph analytics, NLP, deep learning, anomaly detection, reinforcement learning

Business Impact
- Strengthens fraud detection by identifying suspicious behavior more quickly and with greater precision
- Enhances credit decision-making with models that incorporate a broader set of data points
- Improves compliance monitoring by automating the analysis of documents and communications
- Supports better resource allocation by reducing manual review and escalation workload

Industry Adoption

- Major financial institutions are integrating AI into key operational and risk functions. Visa monitors global transactions in real time using AI-driven models to detect fraud patterns. PayPal leverages AI to reduce false declines and improve transaction security, balancing fraud prevention with user experience. JPMorgan Chase uses its COiN platform to process legal contracts at scale—reducing turnaround time from days to seconds.

- Across banking, payments, and capital markets, firms are adopting AI to improve risk posture, optimize cost structures, and comply with regulations more efficiently. Those that integrate AI thoughtfully—balancing automation with transparency and oversight—are seeing measurable gains in resilience, responsiveness, and operational performance.

Operations, Manufacturing, and Supply Chain: *AI-Powered Process Optimization and Predictive Insights*

Core Challenge

Operations leaders face persistent challenges across the value chain: unplanned equipment downtime, inefficient supply chains, and quality control gaps. Traditional forecasting and maintenance tools often fall short in dynamic environments, leading to production delays, stockouts, and costly manual inspections. The pressure to increase output while containing costs and ensuring resilience is mounting—especially as customer expectations rise and global logistics remain volatile.

AI in Action

AI is enabling a new era of responsive operations. Predictive maintenance algorithms monitor equipment health using real-time sensor data to forecast failures before they happen. In manufacturing, computer vision systems conduct automated visual inspections at scale, catching defects invisible to human inspectors. AI-based process controls adjust parameters on the fly to optimize yield, energy consumption, or throughput. Across supply chains, AI enhances forecasting, inventory optimization, and logistics, helping businesses respond to disruption and reduce waste. DHL's adoption of AI-driven logistics, as highlighted by Emerj (2020), illustrates real-world gains in accuracy and responsiveness.

Key Technologies

ML, predictive analytics, digital twins, computer vision, reinforcement learning, generative design, optimization algorithms

Business Impact

- Predictive maintenance cuts breakdowns and extends equipment life.
- Computer vision improves quality and reduces manual inspection overhead.
- AI-enhanced supply chains reduce inventory costs and improve service levels.
- Optimizing production settings in real time lowers energy usage.

Industry Adoption

- AI is increasingly embedded in manufacturing and logistics environments. Siemens applies AI across factories to anticipate equipment failures, reducing unplanned downtime. At BMW's Munich plant, an AI-powered visual inspection system evaluates paint quality more accurately than human inspectors.

- Amazon uses ML for predictive inventory stocking, enabling same-day delivery and reducing fulfillment costs. Shell leverages AI on offshore rigs to optimize maintenance.

These examples illustrate how AI helps organizations unlock new levels of efficiency, resilience, and quality in complex operations.

Human Resources and Talent Management: *Data-Driven People Decisions*

Core Challenge

HR departments face growing pressure to manage talent more strategically, but are often hampered by labor-intensive workflows, subjective evaluations, and difficulty managing scale. Screening thousands of resumes, predicting employee churn, or planning internal promotions remain time-consuming and prone to bias. Organizations struggle to understand employee sentiment and systematically match skills to roles. At the same time, employee expectations around career growth, personalization, and flexibility are rising. Traditional HR tools and methods are ill-equipped to keep up with these demands.

AI in Action

AI brings automation and analytics to talent processes—turning HR into a more data-driven, strategic function. AI tools can instantly screen resumes, analyze candidate profiles, and assess fit, dramatically reducing time-to-hire. NLP-enabled video interview platforms evaluate tone and word choice to gauge communication and soft skills. AI-powered chatbots respond to routine HR inquiries in real-time, improving employee experience while reducing manual load. This shift allows HR teams to focus less on administration and more on strategy, retention, and workforce planning.

Key Technologies

NLP (for resume parsing and chatbot support), ML classifiers (for candidate ranking), sentiment analysis (from surveys and internal data). Bias detection and correction techniques are especially important to ensure fairness in AI-driven HR systems.

Business Impact

- Reduces recruiter screening time

- Lowers the cost of hiring

- Enables faster hiring decisions

Industry Adoption

- Unilever reengineered its high-volume hiring process using AI tools resulting in reduced time to hire and a reduction in recruiter hours invested. IBM uses its Watson Candidate Assistant to analyze resumes against job descriptions, instantly identifying the top applicants. This has helped IBM speed up hiring and improve candidate-job matching.

- More broadly, AI is being used across HR for performance tracking, internal mobility, learning and development, and retention risk prediction. As AI takes over administrative load, HR leaders are turning their attention to more strategic concerns—like culture, engagement, and internal talent mobility.

- To be successful, ethical use of AI in HR remains a priority: you need to embed fairness audits and explainability checks into AI tools to prevent bias.

Healthcare and Life Sciences: *Diagnosis, Discovery, and Personalized Care*

Core Challenge

Healthcare faces high-stakes problems: diagnostic errors, slow drug development, and generalized treatments that often miss individual needs. Clinicians are under pressure to detect diseases earlier, make faster treatment decisions, and reduce costs while navigating complex data environments. Meanwhile, drug development takes years and billions of dollars, and patient outcomes are inconsistent because care isn't always personalized.

AI in Action

AI is transforming healthcare through early detection, accelerated research, and tailored care. Deloitte Consulting (2022) reports that healthcare leaders increasingly consider AI to be a mission-critical tool for improving outcomes and operational efficiency. Computer vision models read X-rays, MRIs, and pathology slides to detect conditions such as cancer or diabetic eye disease earlier than human review. ML helps hospitals flag patients at risk of readmission or complications, enabling preemptive care. In research, AI speeds up drug discovery by modeling protein structures or screening billions of molecules in silico. Precision medicine tools use genetic and clinical data to recommend individualized treatments.

Key Technologies
Deep learning (for image diagnostics), ML (risk scoring and treatment prediction), NLP (to analyze medical records), reinforcement learning (treatment optimization), Gen AI (molecule generation, synthetic data for trials)

Business Impact
- Reduces diagnostic errors and improves early detection rates

- Accelerates drug discovery and clinical trial success

- Cuts administrative costs and boosts workforce productivity

- Enables more personalized, effective treatments

Industry Adoption
- Google's DeepMind developed an AI model for retinal scans that matches expert-level accuracy, helping detect diabetic eye disease early.

- NYU Langone Health uses predictive AI to identify at-risk patients before readmission, improving care coordination and lowering costs.

- Exscientia, a UK-based startup, used AI to design a novel OCD drug in under 12 months—a process that traditionally takes years.

- Hospitals such as UCLA Health and the Mayo Clinic are using AI to automate appointment scheduling and reduce physician paperwork.

AI in healthcare is powerful but not plug-and-play. It requires validated accuracy, clinical oversight, and careful deployment. When implemented thoughtfully, however, it delivers life-saving benefits and operational efficiencies, marking a major evolution in how care is delivered.

AI-Specific Adoption Challenges

Adopting AI is not like adopting traditional technology. AI introduces a new class of challenges that many leaders may not anticipate—issues that stem from its probabilistic nature, reliance on dynamic data, and deep integration into decision-making. This list focuses on challenges that are specific to AI and that require a shift in mindset, governance, or capability to address effectively. The chapters that follow explore these topics in more detail.

Hallucination: *When AI Makes Up Information*
Gen AI systems can produce fluent and convincing, but false information. These "hallucinations" are not bugs; they are a natural consequence of how language models generate outputs from statistical patterns, not facts. This risk is especially concerning in high-stakes domains like law, healthcare, and finance.

Where addressed in this book: Chapter 5, "AI Literacy," p. 103, and Chapter 6, "Risk Level as a Selection Criteria," p. 127

Data Quality and Fragmentation
AI performance is only as good as the data it learns from. If that data is outdated, siloed, biased, or of low quality, model outputs will be flawed regardless of the algorithm. Most enterprises are not yet data-ready for AI at scale.

Where addressed in this book: Chapter 5, "Data Management and Availability," p. 110

Integration into Legacy Systems

AI often fails not because the models are wrong, but because they cannot be embedded into real-world environments. Many organizations underestimate the complexity of connecting AI to existing tech stacks, workflows, and monitoring systems.

Where addressed in this book: Chapter 2, "Integration with Current Technology Stack," p. 60

Compute and Infrastructure Requirements

AI, especially generative models, requires significant computing power and scalable infrastructure. Organizations may be caught off guard by the cost, energy demands, and latency risks of production AI.

Where addressed in this book: Chapter 5, "Technical Infrastructure," p. 108

Bias and Unintended Discrimination

When trained on biased data, AI can reinforce or even amplify existing inequalities. This is particularly dangerous in areas such as hiring, lending, or law enforcement in which fairness is legally and ethically required.

Where addressed in this book: Chapter 2, "Data Governance for Responsible AI," p. 48

Regulatory and Privacy Complexity

AI systems often process personal or sensitive data, and many jurisdictions are introducing regulations that go beyond traditional privacy requirements. Staying compliant is no longer optional.

Where addressed in this book: Chapter 7, "AI Governance, and Legal and Ethical Compliance," p. 137

Trust and Transparency

AI systems, especially deep learning models, are often hard to interpret. If business users and customers don't understand how decisions are made, trust erodes and adoption stalls.

Where addressed in this book: Chapter 2, "Data Governance for Responsible AI," p. 48

These challenges are real, but they are manageable. The rest of this book is designed to help you address them through structured frameworks, case studies, and practical guidance tailored for decision-makers leading AI adoption at scale.

Chapter Summary

AI adoption presents significant opportunities for businesses to enhance decision-making, optimize operations, and drive innovation. From predictive analytics that improve supply chain efficiency to AI-driven fraud detection that secures financial transactions, the technology delivers measurable impact across industries.

However, realizing this impact requires more than deploying tools—it demands overcoming core challenges, including data quality and privacy concerns, integration with legacy systems, talent shortages, and ethical risks. Success depends on taking a strategic, organization-wide approach that aligns AI initiatives with business objectives while managing risk and change.

Actionable Takeaway

Analyze Industry Trends and AI Use Cases—Identify how leading companies in your sector are using AI to solve real business problems, increase operational efficiency, and build competitive advantage. Use these insights to guide early investments and set realistic expectations.

Next Chapter

We have examined how AI is reshaping industries and creating new possibilities. The next step is understanding how AI actually functions beneath the surface. In the following chapter, we will explore the core mechanics of AI and what leaders must grasp to harness its full potential.

Chapter 2

How AI Works
and What Leaders Need to Know

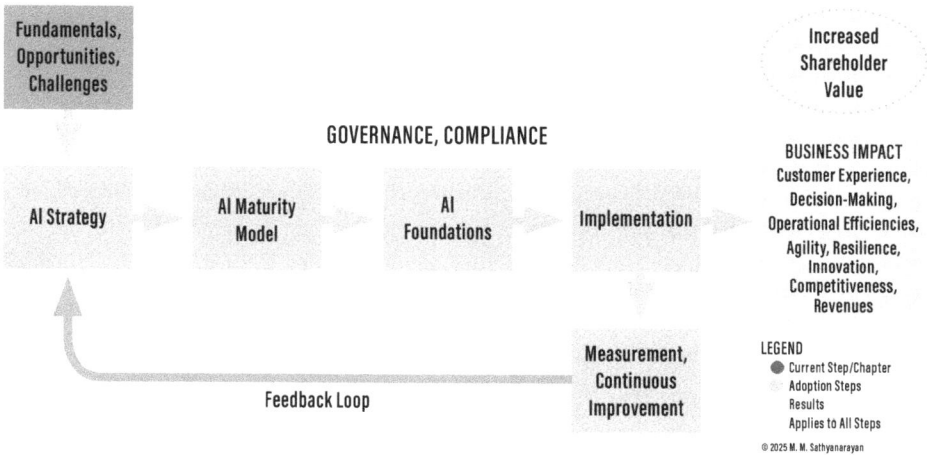

Fundamentals, Opportunities, Challenges

Increased Shareholder Value

GOVERNANCE, COMPLIANCE

AI Strategy → **AI Maturity Model** → **AI Foundations** → **Implementation**

BUSINESS IMPACT
Customer Experience,
Decision-Making,
Operational Efficiencies,
Agility, Resilience,
Innovation,
Competitiveness,
Revenues

Measurement, Continuous Improvement

Feedback Loop

LEGEND
● Current Step/Chapter
Adoption Steps
Results
Applies to All Steps

© 2025 M. M. Sathyanarayan

FIGURE 2.1 AI ADOPTION ROADMAP

Y ou have seen the possibilities and the challenges of AI adoption. Before you begin shaping a strategy for your organization, it is important to understand how AI systems differ from traditional technologies: not at the level of code or architecture, but in ways that directly affect how you lead, allocate resources, and manage risk (see Figure 2.1). This high-level perspective will help you guide discussions with partners, vendors, and technical teams; ask sharper questions; and support your board in making informed, future-ready decisions. What follows is a practical, business-oriented look at how AI systems learn, what technologies power them, and why governance matters more than ever.

How AI Systems Work:
Learning from Historical Data

Unlike traditional programming, which relies on predefined rules, AI systems learn from historical data to recognize patterns and make decisions. Think of training a new employee: instead of following a rigid rulebook, they analyze past invoices, noting how late payments were flagged or discounts applied. Over time, they develop an understanding of patterns, allowing them to process new invoices confidently. Similarly, AI generalizes patterns from data to handle new situations, enabling automation and decision-making without explicit instructions.

This is essentially how AI systems work. The training process involves:

Feeding the AI with historical data: this could be transaction records, customer feedback, or even images, depending on what the AI is being trained for.

Identifying patterns in the data: during training, the AI learns relationships, trends, and behaviors from the data. This learning process is what creates the AI's model, a set of patterns or "rules" that it has inferred from the training data.

Example: Fraud Detection

Let's say you're training an AI system to detect fraudulent transactions. Historical data might include millions of previous transactions, with some flagged as fraud and others as legitimate. The AI studies this data and learns patterns that are common in fraudulent transactions, such as:

- Large purchases made in quick succession
- Transactions from unexpected locations
- Unusual patterns in spending behavior

Once the training is complete, AI now has a model—a tool it uses to evaluate future transactions. When a new transaction occurs, AI applies the patterns it has learned and predicts whether the transaction is likely fraudulent or not. The decision is not made from scratch every time; it is based on the patterns the AI has already learned from historical data.

Learning from Data versus Explicit Programming

This approach of learning from data is fundamentally different from traditional programming:

Explicit programming: You would write rules like, "If a transaction is over $5,000 and made outside the user's home country, flag it as fraud."

AI learning from data: The system observes many examples of fraud and legitimate behavior, and it discovers its own patterns; patterns that might include complex interactions among transaction amount, location, time, and user behavior.

The key advantage is that AI can handle complex, changing scenarios that would be impractical or impossible to explicitly program. For instance, fraud patterns evolve over time. AI systems can adapt by retraining with new data, continually refining their models.

How AI Applies What It Has Learned

Once trained, AI systems operate by applying their learned patterns to new situations. Think of it as the employee from our analogy, now confidently processing new invoices. They look at a new transaction, compare it to the patterns they have learned, and decide how to act.

For example:

> A customer asks a chatbot for help with their bill. The chatbot applies patterns from its training to recognize the question, retrieve relevant information, and craft a response.

> A recommendation system suggests products by comparing your browsing history to patterns of other users who have browsed similar items.

AI does not memorize; it generalizes. That is why it can handle completely new inputs that were not part of its training data. The strength of AI lies in its ability to learn, adapt, and apply patterns from data to real-world tasks.

When we talk about "AI," we are really referring to a combination underlying technologies working together to deliver the unique outcomes AI is known for. The next section provides a functional view of core AI technologies.

Core Technologies Driving AI

AI relies on interconnected technologies, such as Gen AI, ML, and NLP (see Figure 2.2). Each of these addresses specific challenges while collectively enabling transformative outcomes, such as personalized recommendations, automation, and advanced decision-making.

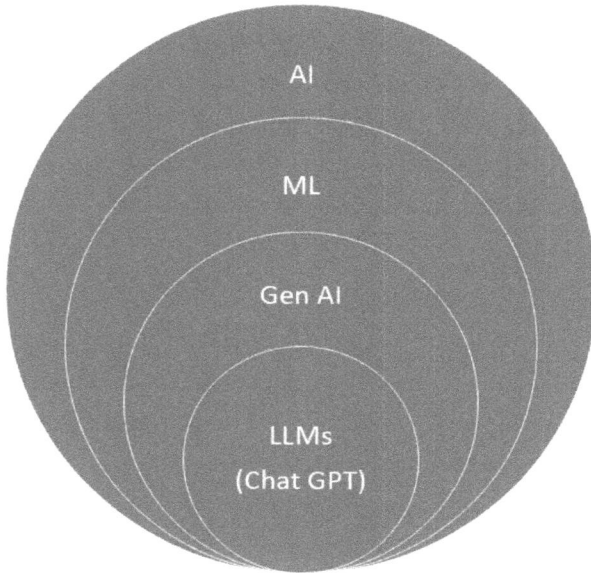

Figure 2.2 Overview of AI Technologies

- **ML: The Cornerstone**

  ML is a core discipline within AI that enables systems to learn patterns and relationships from data, improving their performance over time without explicit programming. By analyzing historical data, ML models can identify trends, make predictions, and drive decisions in dynamic environments. Its adaptive nature allows ML to power applications like recommendation systems, fraud detection, and predictive analytics, continually evolving based on new information.

- **NLP: Unlocking Human Language for Computers**

  NLP is a technology that enables computers to understand, respond to, and communicate using human language, just as we naturally speak or write. It allows systems to make sense of text or speech, such as answering questions, analyzing customer feedback, or translating languages. This makes interactions with technology feel more natural and accessible, improving communication between people and machines in everyday tasks and business applications.

- **Generative AI**
  Gen AI is a specialized class of AI that goes beyond predicting known outcomes to actually creating entirely new data, ideas, or content. Unlike traditional ML systems, which are typically trained to forecast specific results within a restricted set of categories—such as whether a transaction is fraudulent or how much inventory to stock—generative models produce fresh outputs by predicting one piece at a time, continually building on what they've generated so far.

  To see the difference clearly, consider how these individual systems handle the same practical scenario.

  A customer sends an email:

  "Can you explain how your products comply with EU sustainability standards?"

  ### Traditional IT systems (rules-based)
  These systems rely on static, hard-coded rules or keyword triggers. For example, if an email mentions "invoice" or "payment," it might be routed to Billing. If it contains "installation" or "error," it goes to Technical Support. But for a question about "EU sustainability standards," there may be no matching rule, causing the system to either

misroute it to an unrelated queue or default to a general inbox for manual triage.

Traditional ML systems

An ML classifier trained on thousands of labeled customer emails looks for statistical patterns that match its known categories, such as Billing Issue, Technical Support, or Upgrade Request. When faced with this new email, it might produce:

> **Predicted Category: Technical Support (87% confidence)**
> It simply assigns the message to the closest known bucket, even if it's not a perfect fit. It does not generate a response itself. Any reply still depends on downstream business logic or human agents.

Gen AI/LLMs

An LLM, by contrast, doesn't try to classify the message into a predefined category. Instead, it processes the text and begins generating a tailored, natural language response word by word, drawing on vast patterns it learned from its training data. For this same inquiry, it might produce:

"Yes, our products comply with the EU's RoHS and REACH directives, and we maintain ISO 14001 certification for environmental management. I'd be happy to send you our compliance documentation."

There's no need for a routing step or predefined label—the model itself composes a detailed, contextually relevant reply.

The risk of hallucinations

This same mechanism also explains why Gen AI systems can sometimes "hallucinate." Because they predict each next word based purely on statistical likelihood, without directly verifying against factual ground truth, they may occasionally produce information that sounds authoritative, but is inaccurate or entirely fabricated.

Bound by data

It's important to remember that both traditional ML and Gen AI ultimately operate within the boundaries of the data they are trained on. Whether predicting a fixed category or composing a new response, each is constrained—though in very different ways—by the patterns, concepts, and biases embedded in its training datasets. In this sense, data remains the ultimate driver of both capability and risk.

Gen AI in broader applications

Gen AI includes a wide range of systems, from LLMs like Google's Gemini, OpenAI's ChatGPT, and Anthropic's Claude, which generate human-like text and dialogue, to image generators such as DALL•E that create realistic visuals from prompts as well as tools that compose music, design products, or synthesize test data. This streaming, probabilistic nature of prediction is what enables Gen AI to draft new marketing campaigns on the fly, produce software prototypes by learning from extensive code repositories, or create synthetic images that improve quality control in manufacturing.

By harnessing this unique capability to extend learned patterns into entirely new creations, Gen AI unlocks transformative applications across industries. These innovations aren't just about efficiency—they represent fundamentally new ways of turning raw data into competitive advantage.

LLMs: Advanced Tools for Language Understanding
Building on the generative process of streaming predictions step by step, LLMs are a specialized type of Gen AI designed specifically for working with text. Models like ChatGPT are highly capable of tasks such as answering questions; summarizing information; and creating well-structured, coherent content. They achieve this by understanding the context, meaning, and flow of human language, making them essential for applications that require advanced text processing and generation.

Table 2.1 provides an overview of different AI technology components, what they are, their function and examples of how they are used.

Table 2.1 How AI Components Work Together

Aspect	ML	NLP	Gen AI	LLMs
Definition	A field of AI that enables systems to learn patterns from data and make predictions or decisions without explicit programming.	A subfield of AI that focuses on enabling machines to understand, interpret, and generate human language in a natural way.	A subset of AI that creates new, original content (e.g., text, images) by learning from patterns in data.	A specialized form of Gen AI that generates human-like text — and increasingly multimodal outputs — at scale.
Primary Function	Learning and improving based on data to recognize patterns and make predictions or classifications.	Understanding and processing natural language to enable intuitive communication between humans and machines.	Synthesizing new data or content, enabling creative and generative applications.	Enhancing text-based Gen AI with massive training datasets and advanced architectures for nuanced understanding.
Focus	General-purpose pattern recognition and learning across a variety of domains.	Bridging human communication and machine understanding, focusing on language.	Creating outputs that mimic human creativity, such as generating new text, images, or designs.	Mastering language-specific tasks, including summarization, text generation, and contextual understanding.
Output	Predictions, classifications, or decisions based on data.	Structured or analyzed language data.	New and unique outputs that resemble training data (e.g., creative text, art, or media).	Highly refined and human-like text responses tailored to context and input.
Applications	Fraud detection, recommendation systems, forecasting, personalization.	Language translation, chatbots, sentiment analysis, and virtual assistants.	Content creation, design generation, conversational systems, and simulations.	Conversational AI, text summarization, document generation, and complex language understanding.

Let's look at a real-world example. AI is used widely in chatbots. Here is how the components interact:

ML + NLP → Generative AI (LLMs)

- ML provides the foundation for recognizing user intent and extracting patterns.
- NLP ensures natural communication, enabling the chatbot to interpret and respond to user language.
- LLMs — a form of Generative AI — generate dynamic, personalized, and context-aware responses, enabling more sophisticated chatbot interactions.

Example Interaction:
User Query: "Can you check my bill for December and see if I can pay it using my current balance?"

1. ML
   - Identifies the user's intent: "Check bill and balance."
   - Extracts key entities from the query: "December" and "current balance."
2. NLP
   - Breaks the query into actionable components, interpreting it as a request to retrieve specific data (e.g., bill amount and balance).
   - Maintains context for potential follow-up questions, such as "How about January?"

3. Gen AI
   - Constructs a response based on retrieved data:
     "Your December bill is $150, and your current balance is $200, so, yes, you can pay it now."

4. LLM
   - Handles more sophisticated and open-ended queries, such as:
     "Can you compare my December and January bills, and let me know if there are any unusual charges?"
   - Leverages pre-trained knowledge and understanding of billing terminology to analyze the bills and generate a detailed summary.

Together, these technologies create a cohesive, intelligent chatbot capable of understanding, responding, and acting effectively in various scenarios.

These technologies depend on good quality data. AI performance is only as good as the data it learns from and that is covered in the next segment.

How Data Shapes AI Performance

In this section, we'll explore different types of data AI relies on (see Figure 2.3), the lifecycle of data in AI systems, and the challenges organizations face when managing data for AI. Finally, we will discuss the role of data governance in ensuring ethical, scalable AI implementations.

Types of Data in AI

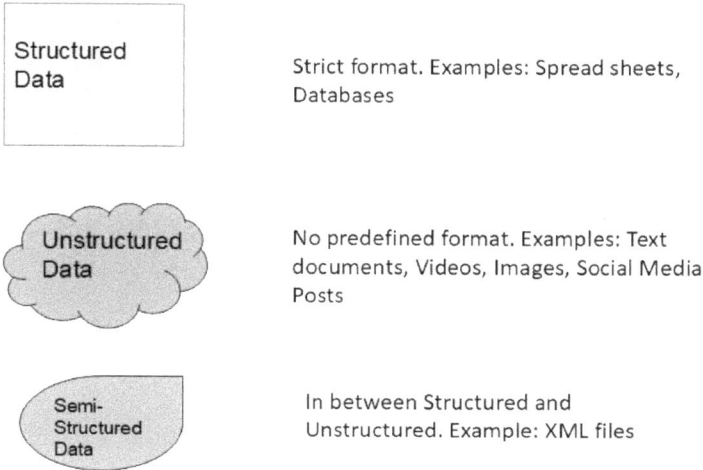

Structured
Data

Strict format. Examples: Spread sheets,
Databases

Unstructured
Data

No predefined format. Examples: Text
documents, Videos, Images, Social Media
Posts

Semi-
Structured
Data

In between Structured and
Unstructured. Example: XML files

Figure 2.3 Types of Data

AI systems rely on three primary types of data, each serving unique roles:

1. Structured Data
   - Definition: Highly organized, stored in rows and columns, such as in databases or spreadsheets
   - Examples: Financial records, sales transactions, customer information
   - AI Applications: Structured data powers tasks like fraud detection, forecasting, and supply chain optimization

2. Unstructured Data
   - Definition: Raw, unorganized data without a predefined format

- Examples: Social media posts, images, videos, audio recordings
- AI Applications: Unstructured data fuels applications like sentiment analysis, image recognition, and NLP

3. Semi-Structured Data
   - Definition: A mix of structured and unstructured data, often organized in hierarchical or flexible formats
   - Examples: XML files, JSON data
   - AI Applications: Used in web applications and integration tasks, ensuring compatibility across diverse systems

Understanding these data types enables you to align AI technologies with specific organizational needs.

The Data Lifecycle in AI

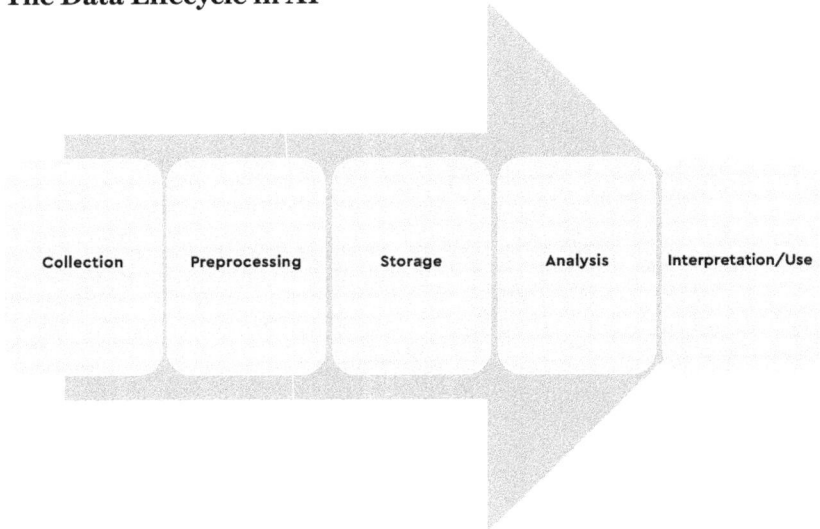

Figure 2.4 Data Lifecycle

The journey from raw data to actionable AI insights follows a structured lifecycle (see Figure 2.4). Each stage plays a critical role in ensuring AI systems deliver accurate and meaningful outcomes:

1. Collection
   - Purpose: Gathering relevant data from internal systems, external sources, or IoT devices
   - Key Considerations: Ensure diversity, relevance, and compliance with privacy regulations during collection

2. Preprocessing
   - Purpose: Cleaning, normalizing, and organizing data to ensure it is ready for analysis
   - Impact: High-quality preprocessing eliminates inconsistencies, reduces noise, and improves AI model performance

3. Storage
   - Purpose: Securely storing data for easy access and analysis
   - Options: On-premises, cloud-based, or hybrid systems, with encryption and redundancy to safeguard integrity

4. Analysis
   - Purpose: Applying algorithms to uncover patterns and generate insights
   - Applications: Predictive modeling, anomaly detection, and trend analysis

5. Interpretation and Use
   - Purpose: Translating insights into actionable decisions or automation
   - Example: Using predictive analytics to personalize customer recommendations or optimize operations

Managing the data lifecycle effectively is key to building reliable, impactful AI systems, but there are challenges you may encounter.

Challenges in Data Management
Data preparation for AI is one of the underestimated aspects of AI adoption, both in terms of cost and time. Organizations often face significant challenges in acquiring, cleaning, and managing data, which can delay projects, increase expenses, and undermine AI's effectiveness if not addressed proactively. Key challenges include:

1. Fragmented Data
   - Description: Data is often siloed across departments or systems, making integration difficult and resource intensive.
   - Cost Impact: Resolving fragmentation typically requires investments in data integration tools and cross-departmental collaboration.
   - Example: A global retailer with customer data stored separately across regional branches delays AI-driven personalization and requires costly efforts to consolidate disparate datasets.

2. Time-Intensive Cleaning
   Preparing data for AI involves resolving inconsistencies, addressing biases, and filling gaps in datasets. Contrary

to a common misconception that "AI can prepare data for AI," the reality is that while AI may accelerate parts of the process, skilled data engineers are still essential to get data in shape before it reaches AI models (TDWI, 2024). This makes data cleaning a labor-intensive task that often consumes a significant portion of AI project budgets, especially in industries with unstructured or poorly maintained data.

Example: Healthcare organizations may spend months cleaning fragmented patient records before deploying AI systems, delaying project timelines and adding operational costs.

3. Compliance and Privacy
   Description: Regulations like GDPR and HIPAA require strict controls over how data is collected, stored, and used, adding complexity to AI initiatives.
   - Cost Impact: Ensuring compliance often requires investments in specialized tools, legal expertise, and governance frameworks. Noncompliance can result in costly fines and reputational damage.
   - Example: Anonymizing sensitive data is crucial for AI initiatives in industries like healthcare and finance, often requiring dedicated teams and advanced technologies to meet regulatory requirements.

A hospital's experience, detailed later in this chapter, underscores that addressing data silos, improving data quality, and meeting anonymization requirements are not optional steps, but essential prerequisites for building reliable and ethical AI systems.

Data Governance for Responsible AI
Strong data governance ensures that AI systems operate transparently, ethically, and in compliance with regulations.

Why Governance Matters
1. Bias Mitigation
   - Regular audits reduce biases in datasets, ensuring fairness in AI outputs.
   - Example: Diverse governance teams, including ethics and domain experts, strengthen oversight.
2. Regulatory Compliance
   - Adhering to privacy frameworks like GDPR and HIPAA minimizes legal risks.
   - Example: Data residency laws may influence where and how data is stored for compliance.
3. Building Trust
   - Transparent AI systems foster confidence among users and stakeholders.
   - Example: Explainability tools, such as dashboards, provide visibility into AI decision-making processes.

Integrating Governance Early
Rather than treating governance as an afterthought, organizations should integrate it into the earliest stages of AI development. This approach ensures:

- Anticipation of Regulatory Changes: Staying ahead of emerging legislation like the EU AI Act

- Reputation Management: Avoiding damage caused by biased or unethical decisions

- Cross-Functional Collaboration: Aligning legal, technical, and business teams to establish a unified governance framework

While we have touched on governance specific to data in this segment, we go into more details of broad governance issues in Chapter 7.

Case Study: *AI-Powered Hospital Transformation*

Background

A major U.S. hospital system embarked on a multiyear AI initiative to automate hospital operations, improve patient care, and enhance decision-making. With over 30 years of patient data, including doctor notes, lab reports, and administrative records, the hospital sought to leverage AI for predictive analytics, workflow automation, and operational efficiency.

However, the project faced several challenges:

- Privacy and Compliance—Patient data contained protected health information (PHI), requiring extensive preprocessing and anonymization before it could be used for AI applications.

- Data Complexity—Medical records were a mix of structured data (age, gender, lab results) and unstructured data (doctor notes, prescriptions), requiring significant data organization and cleaning before model training.

Choosing the Right AI Model

The hospital initially implemented traditional AI models, expecting them to be sufficient. However, it soon realized these models were too rigid, leading the hospital to transition to LLMs such as LLaMA 2—an LLM similar to ChatGPT but developed by Meta (formerly Facebook). While it doesn't learn autonomously in real time, the model can be regularly updated with new data, making it more adaptable than traditional systems.

What Is LLaMA and Why Did the Hospital Use It?

LLaMA (Large Language Model Meta AI) is a family of advanced AI models designed for natural language understanding and generation. Unlike traditional AI models, which rely on predefined rules and structured data, LLaMA can process free-text information, such as doctor notes and medical reports, to extract insights.

The hospital turned to LLaMA 2 (a 7B parameter model) to address a major limitation in its AI system—the inability of traditional models to process unstructured medical data. By integrating LLaMA, the hospital was able to unlock deeper clinical insights, improve predictions, and enhance decision-making.

Starting with Traditional AI for Patient Stay Predictions

One of the hospital's key objectives was to predict patient length of stay to assist with resource planning and capacity management. The initial approach used traditional AI models, which relied solely on structured patient data such as age, gender, and past medical history.

What the hospital did:

- Built a predictive AI model trained on historical patient records.

- Used structured data fields like diagnoses, medications, and demographics to estimate hospital stay duration.

- Applied rule-based thresholds to flag high-risk patients who might need extended care.

Challenges with the initial model:

- Predictions became less accurate over time because the model was not learning from new patient records.

- Manual updates were required whenever new treatments or protocols were introduced.

- The model ignored critical insights found in unstructured data like doctor notes and test results.

This approach provided useful initial insights but lacked adaptability, leading the hospital to explore more advanced AI models.

Transitioning to LLMs for Context-Aware Insights
To improve accuracy and capture richer medical insights, the hospital transitioned to LLaMA 2, an LLM capable of processing both structured and unstructured patient data.

What changed:

- LLMs were fine-tuned with 30 years of hospital data, including doctor notes, lab reports, and treatment histories.

- The new AI model could process free-text medical notes, allowing for more precise and context-aware predictions.

- AI-generated reports provided doctors with summarized patient histories, improving diagnostic accuracy.

This transition was not just about upgrading the AI model—it was about shifting the hospital's entire approach to AI. Instead of solving isolated problems, AI was now reshaping how entire functions operated—from diagnostic decision-making to hospital resource management.

Overcoming the Computational Bottleneck

One unexpected challenge was the significant computing power required to train and fine-tune LLaMA 2.

What they encountered:

- Fine-tuning LLaMA 2 required high-end hardware, cloud-based AI infrastructure, and ongoing resource costs.

- Training a fine-tuned 70B parameter model proved impractical due to high costs.

- The hospital optimized the AI model by using a smaller version (LLaMA 2 7B) while selectively fine-tuning parts of the dataset.

The lesson for leaders: the power of AI comes with significant cost consideration. Choosing the right model is not just about accuracy, but also sustainability and affordability.

Bringing Doctors on Board: Overcoming Resistance to AI

While AI adoption promised efficiency gains, doctors initially saw it as a threat to their expertise. The hospital had to shift perceptions from replacement to augmentation.

How they gained doctor cooperation:

- Demonstrated how AI-powered radiology models reduced diagnostic workload, allowing doctors to focus on complex cases rather than routine scans.

- Showed that AI clinical decision support systems provided evidence-based recommendations rather than overriding physician judgment.

- Introduced a chatbot assistant that retrieved medical records faster, allowing doctors to spend less time on documentation and more time with patients.

Through these efforts, the AI system was positioned as a tool to enhance—not replace—medical expertise.

AI adoption requires more than technical deployment, it demands proactive change management across functions. Leaders must anticipate resistance and frame AI as an enabler of expertise rather than a replacement for human roles. We will provide more insights about change management in Chapter 5.

Expanding AI Beyond Predictions: *Automating Hospital Operations*

With predictive models running effectively, the hospital expanded AI into operational automation.

What the hospital automated:

- AI-assisted discharge planning—predicting optimal discharge dates and freeing up hospital beds faster

- Billing and administrative workflows—Reducing manual paperwork by automating insurance and medical record processing

- AI-driven chatbots—Helping nurses and administrative staff quickly access patient information

This approach transformed AI from a prediction tool into a fully integrated operational system.

Lessons Learned: AI as a Function-Wide Transformation
This hospital's AI journey highlights key considerations for business leaders evaluating AI solutions:

1. AI Can Transform Entire Business Functions—Not Just Solve Isolated Problems.
   - The hospital's AI journey started as a predictive tool for length of stay, but it evolved into an AI-powered automation system affecting hospital-wide functions.
   - Leaders should think beyond point solutions and consider how AI can reimagine entire functions like marketing, software development, and customer service.

2. Choosing Between Traditional AI and LLM-Based AI
   - Traditional AI models are less expensive, but require manual retraining and struggle with unstructured data.
   - LLM-based AI is more powerful, but requires significant computing power and higher costs.

3. The Hidden Cost of AI Implementation
   - The hospital faced high computational costs in fine-tuning models like LLaMA 2 (7B and 70B).
   - Leaders must factor in hardware, cloud infrastructure, and ongoing AI training expenses when deciding among different AI approaches.

4. The Importance of Data Quality
   - Before AI could provide value, the hospital had to spend months cleaning and structuring data.

5. Adoption Resistance and Organizational Change
   - Some doctors initially saw AI as a threat to their expertise.
   - AI should be positioned as a decision-support tool rather than a replacement for human judgment.

Leaders must carefully balance AI's transformative potential with cost, complexity, and long-term sustainability. While advanced AI models can drive efficiency and unlock new capabilities, they also require significant investment in infrastructure, talent, and ongoing maintenance. Success depends on aligning AI adoption with clear business objectives, ensuring that AI initiatives not only deliver value, but are also financially and operationally viable over time.

So far, we have covered how AI systems work, the core technology, and the importance of data, and we presented a real-world case study along with pointers for leaders. In addition, you need to be aware of how AI development is different, its implication on cost and schedule, and the implications of integrating with your current technology stack. These are covered in the next two segments.

AI versus Traditional Development:
What Leaders Must Understand

AI and data science projects differ significantly from traditional software development, requiring a shift in mindset, tools, and processes (see Table 2.2). While traditional development follows structured processes with clearly defined outcomes, AI projects are exploratory and iterative, relying on experimentation with data to uncover patterns, insights, and relationships. These differences make managing AI projects challenging (see Figure 2.5).

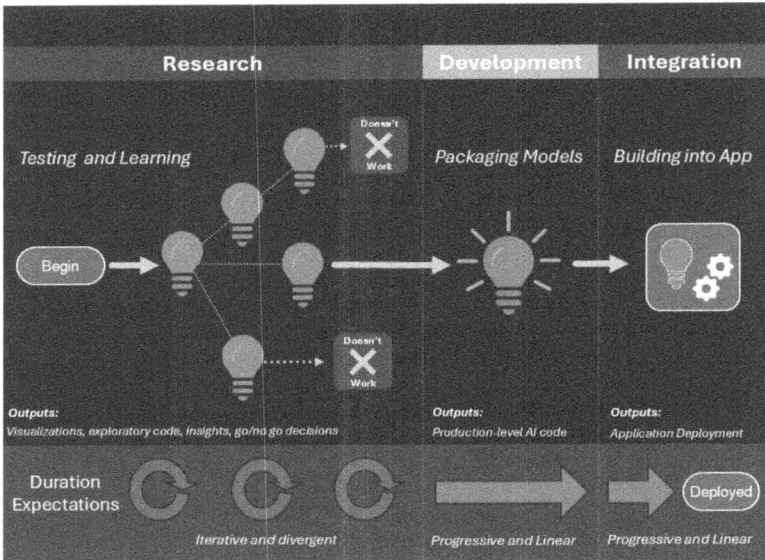

Figure 2.5 AI Development Lifecycle

"The processes to study and build an algorithm or any 'AI' technology have many similarities to what is found in a standard software engineering framework. However, where Data Science is distinct, the differences are noteworthy. Data science has code artifacts similar to building any application; however, there are many other 'exploratory' steps that are completed prior to building anything that would be considered 'useful'. The distinctive Data Science steps in this process are referred to as 'research' in this document and are conducted using an exploratory scientific process."

Source: *Data Science Processes and Standards*

Ion Nemteanu, Nemsee Consulting,

Rady School of Management, UCSD

Table 2.2 Key Differences

Aspect	AI/Data Science Projects	Traditional Development
Nature of Work	Exploratory, involving research and experimentation	Defined and structured, following clear requirements
Process	Iterative, driven by scientific methods and hypothesis testing	Engineering-focused, often using Agile or Waterfall approaches
Outcome	Insights, predictive models, and evolving solutions	Static, predefined software applications
Tools	Statistical analysis, ML frameworks, data visualization	Programming languages, development frameworks

AI Development

AI is fundamentally different from traditional technology initiatives. Instead of starting from a fixed blueprint, teams explore data-driven questions, run experiments to uncover patterns, and refine models through cycles of learning. This process typically involves:

- Generating and testing hypotheses to discover trends or predict outcomes

- Rapidly experimenting using tools like Jupyter, TensorFlow, and PyTorch
- Continuously adjusting models to improve accuracy and business relevance

Strategic implications for leadership

Because of this exploratory nature, managing AI requires a shift in how leaders think about timing, investments, and oversight:

- Flexible but governed plans
  AI timelines and budgets often evolve as new insights emerge. This does not mean a blank check. It means setting expectations that resources may shift—when justified by strong business cases—under clear governance.

- Responsiveness to discovery
  As data reveals new opportunities or risks, priorities may need to change. Formal checkpoints ensure leadership decides when to pivot, scale up, or stop, keeping efforts aligned with enterprise strategy.

- Transparent communication
  Given less predictable outcomes, regular updates on progress, risks, and costs are essential to keep stakeholders informed and engaged.

- Balancing risk with long-term advantage
  AI usually requires higher upfront costs and patience for returns. Leaders must weigh this against the potential to build capabilities competitors can't easily replicate.

Case in Point: AI-Powered Hospital Transformation (covered earlier)

This U.S. hospital illustrates exactly how these challenges were managed. It began by trying to predict patient length of stay using traditional AI models trained on structured data like diagnoses and demographics. This delivered initial value—helping with resource planning—but couldn't keep pace as treatments evolved or new data types emerged.

When predictions degraded and manual retraining became unsustainable, the hospital pivoted. Leaders approved an investment to transition to LLMs like LLaMA 2, which could process free-text doctor notes and lab reports. This shift wasn't planned at the outset; it was a strategic course correction based on what the data revealed. They:

- Re-scoped timelines and budgets to accommodate the new model and the intensive data cleaning needed

- Managed high infrastructure costs by optimizing to a smaller LLaMA variant, aligning with financial constraints while still enhancing capability

- Engaged physicians early, positioning AI as a diagnostic assistant—reducing resistance and ensuring adoption

- Established clear governance to decide which insights justified new investments, keeping control firmly with leadership

This evolution turned what began as a point solution into a function-wide transformation. AI now informs diagnostics, discharge planning, billing, and more—all because leadership set the expectation that plans could adapt, but only under rigorous review and with a clear link to strategic value.

Integration with Current Technology Stack

A key question that comes up while discussing AI adoption is how it works with current technology stack. After all, you still need the traditional IT systems.

In many cases, AI works in conjunction with other technologies such as robotic process automation (RPA), the Internet of Things (IoT), and conventional enterprise systems to create a more powerful and integrated solution.

For example, invoice processing involves structured data like purchase orders and payment records. However, invoices often arrive as PDFs or scanned documents containing unstructured text, making automation challenging. Instead of relying on staff to manually extract and enter details, AI can leverage NLP and optical character recognition (OCR) to convert unstructured text into structured data. Once processed, RPA handles validation, record updates, and approvals, minimizing human effort and errors.

Chapter Summary

This chapter introduced how AI systems learn from data and why they behave differently from traditional software. It covered the core technologies driving modern AI—from ML to LLMs—and how they combine to deliver business value. A real-world hospital case study illustrated how AI can evolve from pilot to platform, and why data quality, governance, and development practices are critical leadership concerns when scaling AI.

Actionable Takeaway

Use your understanding from Chapters 1 and 2 to bring your leadership team up to speed. For your organization to build an effective AI strategy, your key stakeholders need a common understanding of how AI works, where it adds value, and what challenges to prepare for.

Next Chapter

You now have a foundational view of AI's capabilities, limitations, and underlying technologies. In the next chapter, we will begin shaping an AI strategy aligned with your organization's goals and operating environment.

Chapter 3

Developing AI Strategy

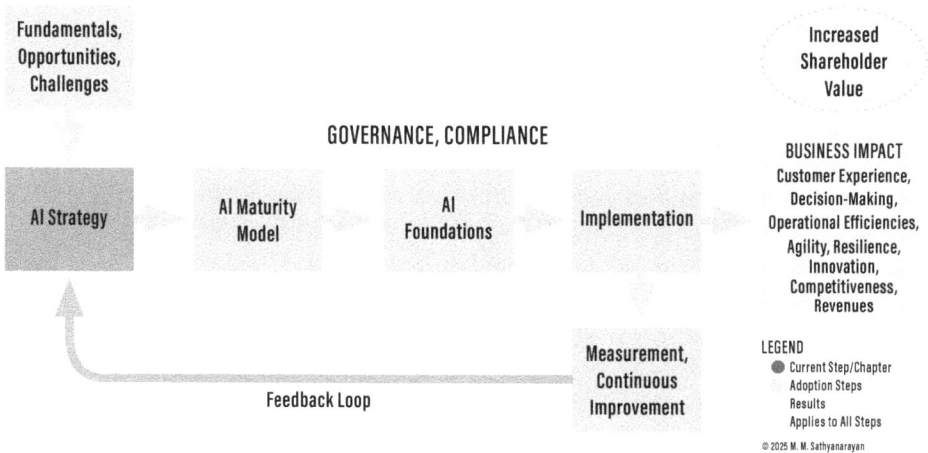

Fundamentals, Opportunities, Challenges

Increased Shareholder Value

GOVERNANCE, COMPLIANCE

AI Strategy → AI Maturity Model → AI Foundations → Implementation

BUSINESS IMPACT
Customer Experience, Decision-Making, Operational Efficiencies, Agility, Resilience, Innovation, Competitiveness, Revenues

Measurement, Continuous Improvement

Feedback Loop

LEGEND
● Current Step/Chapter
Adoption Steps
Results
Applies to All Steps

© 2025 M. M. Sathyanarayan

FIGURE 3.1 AI ADOPTION ROADMAP

We have now covered what AI is, how it creates value, and the challenges it introduces. This chapter moves into the next stage of the AI Adoption Roadmap: developing a strategy aligned with business goals. It introduces seven guiding principles to help you shape an approach that is practical, resilient, and tailored to your organization's context.

Figure 3.1 shows where strategy fits in the AI Adoption Roadmap—linking foundational understanding with enterprise execution. As this chapter will explore, strategy is not a document; it is an adaptable, leadership-driven process for aligning AI with business value.

Guiding Principles for AI Strategy

1. Align AI Initiatives with Business Objectives—Ensure every initiative supports strategic goals.

2. Invest in Leadership and Vision—Equip leaders to champion AI adoption with authority and clarity.

3. Prioritize Capability Building While Demonstrating Value—Balance quick wins with long-term investment.

4. Balance Temporary Staffing, Hiring, Upskilling, and Partnering for AI Talent—Build a sustainable, blended talent strategy.

5. Establish Risk-Aligned and Scalable Governance—Govern AI as a strategic risk management function.

6. Design for Human–AI Collaboration in the Age of Emerging Autonomy.

7. Foster Adaptive AI Strategies—Combine continuous learning with structured iteration.

These principles form the foundation of a durable AI strategy—whether your organization is just starting out or scaling enterprise-wide.

1. Align AI Initiatives with Business Objectives

To maximize impact, AI efforts must be anchored in strategic business goals—while still allowing room for experimentation. In early stages, experimentation can be part of the strategy itself, provided that expectations, investment, and metrics are clearly aligned.

This connection between business value and AI execution ensures initiatives deliver measurable outcomes, such as operational efficiency, enhanced customer experiences, product innovation, revenue growth, or structured exploration of emerging capabilities.

Leadership Actions

- Define clear business goals.
  Clarify the objectives driving your AI investments—whether reducing costs, improving retention, expanding markets, or enabling new capabilities. In some cases, early AI literacy may be a starting point, but it should ultimately support measurable business outcomes.

- Map AI to strategic priorities.
  Identify how AI can directly support those goals.

  Example: Use predictive analytics to optimize supply chains, or deploy chatbots to reduce support load and improve service quality.

- Embed AI into core functions.
  Integrate AI into existing workflows, systems, and decision cycles. Some organizations may drive this top-down, while others evolve through local experimentation. In either case, alignment between strategy and execution is essential.

- Set measurable outcomes.
  Define KPIs that reflect both business and technical success. Business units may track efficiency, growth, or satisfaction, while technical teams measure performance through accuracy, uptime, or system latency. These should roll up to shared outcomes.

Example: Increase customer satisfaction scores by 10% in six months using AI-powered chatbots, while tracking resolution rate, response time, and iteration cycles.

With AI efforts anchored in clear business outcomes, the next step is to ensure the right leadership is in place—equipped, accountable, and empowered to drive execution at scale.

2. Invest in Leadership and Vision

AI success depends on leadership—clear direction, aligned priorities, and sustained commitment. Organizations that designate senior leaders to own AI adoption avoid common pitfalls like fragmented initiatives, duplicated efforts, or misalignment between technical work and business value.

Leadership in AI is not only about technology expertise; it requires strategic fluency, business context, and the ability to influence across functions. DeNittis (2022) highlights that overlooking cross-functional alignment and model integration can lead early AI projects to fail before they scale.

While these qualities are important for any technology-driven transformation, AI introduces distinct challenges that elevate the leadership bar: it spans both technical and ethical domains, evolves rapidly, and carries risk implications that reach the boardroom. This principle is about ensuring the right individuals are equipped, positioned, and empowered to lead with clarity and confidence.

Leadership Actions

- Appoint a senior AI leader.
  Assign a senior leader—such as a Chief AI Officer or Head of AI—with responsibility for driving AI strategy and execution across the enterprise. This ensures that the flexibility, budget adjustments, and responsive decision-making discussed earlier under strategic leadership have a clear owner.

- Position the role strategically.
  Ensure this leader has influence in enterprise decision-making forums and clear authority over budget, talent, and operations.

- Strengthen AI fluency across leadership.
  Use executive briefings, expert advisors, or structured peer sessions to improve organizational understanding of AI's capabilities and risks.

- Align vision with business strategy.
  Collaborate across functions to develop a clear AI vision that reflects strategic goals, technical feasibility, and enterprise values.

- Communicate the vision effectively.
  Reinforce the vision through internal communications, executive forums, and dialogue that links AI work to broader transformation efforts.

With visible, empowered leadership in place, the next priority is building the internal capabilities needed to deliver meaningful, sustained progress.

3. Prioritize Capability Building While Demonstrating Value

AI's transformative potential takes time to realize, but that doesn't mean value must wait. Early projects should deliver tangible outcomes—however modest—while simultaneously building the capabilities required for long-term success. This dual focus builds organizational confidence, earns stakeholder trust, and lays the groundwork for sustainable growth.

Leadership Actions

- Start with strategic pilots.
  Focus early efforts on real business challenges that can yield useful insights and momentum. As Bhaskar Ghosh (2022) advises, selecting automation or AI projects with clear value alignment is key to avoiding wasted investment and inertia. Choose low-risk, high-potential areas that demonstrate AI's value without disrupting core operations.

  Example: Use AI to enhance a customer support function (e.g., chatbot-assisted service), then refine based on real interactions and feedback.

- Establish clear success metrics.
  Define metrics that reflect both business value and capability development. Include tangible outcomes (e.g., cycle time, cost savings) and intangible ones (e.g., collaboration, readiness).

  Example: A chatbot pilot reduces response times by 15%—a measurable win that also builds internal expertise.

- Focus on learning and iteration.
  Treat early projects as structured learning cycles. Capture

lessons from both successes and setbacks, and use them to shape future initiatives and resource allocation.

- Maintain accountability.
  Set milestones that reflect both near-term delivery and progress toward maturity. Hold teams accountable for learning outcomes as well as implementation.

  Example: Deploy a functioning AI model within six months, with success measured by timeline adherence and clarity of lessons learned.

- Evolve metrics as maturity grows.
  Begin with readiness-focused KPIs—such as percentage of trained staff or time-to-deployment—and shift toward production-oriented metrics like ROI or AI-enabled revenue as capabilities scale.

  Example: Aim for a 30% pilot-to-production transition rate early on, increasing to 70% as maturity improves.

With clear metrics, targeted pilots, and a focus on organizational learning, capability building becomes a driver—not a delay—in delivering meaningful AI results.

4. Balance Temporary Staffing, Hiring, Upskilling, and Partnering for AI Talent

Developing AI capabilities requires a flexible, multipath approach to talent. Early-stage initiatives benefit from temporary staffing models that enable experimentation without long-term commitments. As maturity grows, organizations must shift toward a more permanent structure that blends targeted hiring, employee upskilling, and external partnerships. This balanced approach ensures scalability while staying aligned with evolving business needs.

Leadership Actions

- Use temporary staffing to support early experimentation. Engage consultants, offshore resources, or contract-based AI specialists to prototype solutions and refine strategic focus before committing to permanent roles.

  Example: Bring in contract data scientists to develop and test a recommendation engine before deciding on long-term resourcing.

- Identify critical roles for permanent hiring.
  As initiatives move into production, focus hiring efforts on roles that guide implementation, scale use cases, and manage risk.

  Example: Hire a senior AI strategist to oversee cross-functional pilots and align them with enterprise goals.

- Upskill employees across business and technical functions. Build organizational readiness by equipping teams with AI literacy and application skills—tailored to their roles.

  Example: Train marketing teams on predictive segmentation, or enable HR teams to use AI for sourcing and screening.

- Form strategic partnerships to accelerate capability development. Collaborate with universities, training providers, or AI vendors to supplement internal skills and provide hands-on learning opportunities.
  Example: Partner with an AI vendor to co-develop a recommendation system, allowing internal teams to learn through execution.

- Track progress and talent ROI.
  Monitor time-to-fill for critical roles, employee engagement with training, and the value of external partnerships. Adjust resourcing strategy as business needs evolve.

  Chapter 5 offers a deeper look at the roles and capabilities needed to operationalize this talent strategy.

With leadership, early wins, and the right talent model in place, the next step is to embed governance into your strategy, ensuring AI remains ethical, accountable, and aligned with enterprise risk management.

5. Establish Risk-Aligned and Scalable AI Governance

Governance is not simply a compliance task; it is a strategic risk management function. As AI capabilities expand, so do the potential consequences of unmanaged bias, opaque decision-making, and regulatory exposure. Effective governance ensures that AI delivers value while managing legal, ethical, and operational risks at scale.

Embedding governance early in your strategy signals that AI is not a technical experiment—it is a business capability with executive-level implications. The role of governance is to enable innovation responsibly, with trust and transparency built in from the outset.

Leadership Questions

- Where are we currently using AI—and who is accountable for its outcomes?

- Are we focusing governance where risks are highest?
- Do we have visibility into bias, model drift, or unintended impacts?
- Are we prepared for the AI-related regulations emerging in our sector?
- Is our governance structure aligned with our current AI maturity level?
- Are third-party and Gen AI systems governed with the same care as internal solutions?

These questions lay the groundwork for building a governance model that adapts as your AI strategy evolves. Chapter 7 provides a full roadmap, including the Minimum Viable Governance (MVG) framework—a practical, risk-aligned approach that scales oversight as adoption grows.

Addressing governance at the strategy stage ensures your AI initiatives can scale with confidence while meeting rising expectations from customers, regulators, and boards.

6. Design for Human–AI Collaboration in the Age of Emerging Autonomy

AI's role in the enterprise is evolving. In its first wave, AI supported human decisions—analyzing data, automating tasks, and augmenting productivity. But a more profound shift is on the horizon: the rise of the autonomous enterprise, where AI begins to drive decisions, adapt workflows in real time, and operate with minimal human input.

This transition doesn't eliminate the human role; it redefines it. For example, Microsoft 365 Copilot embeds AI into workflows to assist users contextually in real time as Microsoft (n.d.) describes.

Back in 2018, *Harvard Business Review* introduced the concept of "collaborative intelligence," showing that companies outperformed peers when humans and AI worked together in a shared system of strengths. Rather than rendering this idea obsolete, the emergence of more autonomous AI systems confirms its importance. As AI becomes more agentic in behavior—even if not fully autonomous—the human role shifts from executor to supervisor, orchestrator, and ethics governor.

The Daugherty Principle (2018), which states that organizations must reengineer business processes to enable AI's full potential, remains a critical guidepost. But now, the design challenge is more complex: ensuring that AI systems can act, learn, and adapt without losing human oversight, values, or control.

Recent research supports this evolution. According to the Accenture Technology Vision 2024, enterprises are entering an "era of AI teammates," in which collaboration is not just about support, but also about managing distributed agency. The goal isn't to remove people; it's to position them at the center of judgment, governance, and value alignment.

Leadership Actions

- Reengineer workflows for shared agency.
  Identify decision points where AI can act independently and define where human intervention is required. Design escalation, override, and review pathways from the start.

- Build supervisory fluency.
  Train employees—not just to use AI, but to evaluate,

interpret, and course-correct its actions in context. This includes understanding model limitations, bias risks, and edge-case handling.

- Redefine success metrics.
  Move beyond efficiency gains. Track metrics like trust in AI recommendations, frequency of human-AI handoffs, or alignment with strategic goals to assess the quality of collaboration.

- Balance automation with accountability.
  Don't mistake automation for maturity. Ensure every AI-driven process includes clear accountability structures, especially in customer-facing or regulated functions.

- Communicate the strategic shift.
  Frame AI not as a tool, but as an evolving collaborator. Set cultural expectations that humans remain essential to oversight, judgment, and learning.

7. Foster Adaptive AI Strategies

AI adoption is not a one-time implementation; it is a continuous journey shaped by the unique nature of AI itself. Unlike traditional systems that remain static until manually updated, AI models learn and evolve as they process new data and uncover fresh patterns. Each interaction is a chance to refine future outcomes, demanding strategies that are adaptive, iterative, and designed to harness AI's evolving intelligence.

Consider a cookie delivery company aiming to predict which customers will reorder next month. A traditional approach might rely on a fixed rule: if a customer spent over $50 last month, flag them for a promotion. That rule stays the same until someone explicitly changes it—it doesn't improve with experience.

By contrast, an AI model trained on historical data might discover that customers buying chocolate chip cookies on rainy days are highly likely to reorder, while late-night, first-time buyers rarely do. As more orders come in, new behaviors emerge: a marketing campaign could boost oatmeal raisin sales or seasonal trends might shift demand. The AI model doesn't automatically adjust to these changes; it requires new data, evaluation, retraining, and redeployment to stay effective.

This cycle of observing fresh data, orienting by analyzing results, deciding how to adapt, and acting by updating the system mirrors the OODA loop: Observe, Orient, Decide, Act. Embracing this framework ensures organizations build AI strategies that not only leverage continual learning, but systematically evolve over time.

The OODA loop, originally developed for high-stakes decision-making, helps teams synthesize new data, align insights with strategic goals, and execute AI initiatives through responsive, iterative cycles. Figure 3.2 illustrates how continuous learning feeds this loop—enabling informed, adaptive AI execution.

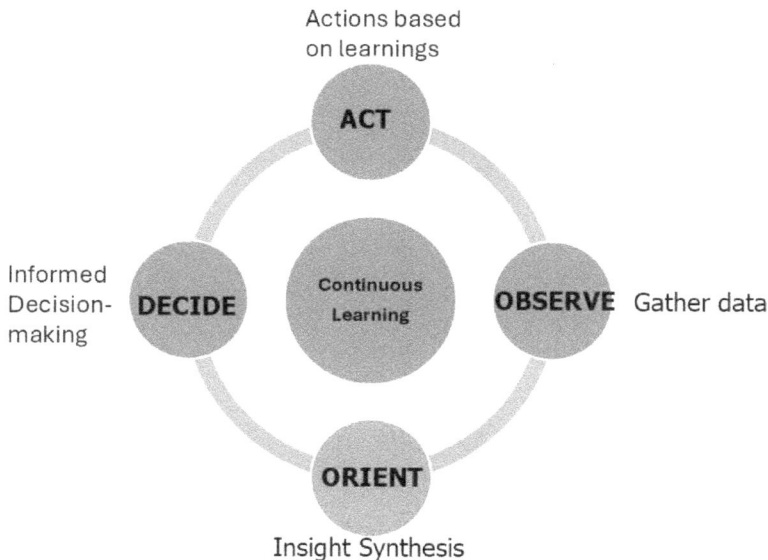

Figure 3.2 The OODA Loop

● Observe → AI teams analyze new data and emerging trends.

● Orient → The teams interpret insights in alignment with business goals.

● Decide → Leadership identifies key AI initiatives based on learning.

● Act → Teams implement and refine AI solutions, feeding new insights into the cycle.

Leadership Actions

- Build structured feedback loops.
  Capture insights from AI initiatives, including performance metrics, user feedback, and operational outcomes. Use this data to refine tools, workflows, and deployment decisions.

  Example: Monitor satisfaction scores for AI chatbots and adjust the model based on patterns in user feedback.

- Encourage cross-functional learning.
  Create opportunities for teams to share experiences, uncover common challenges, and accelerate adoption.

  Example: Host monthly "AI Learn & Share" sessions to surface insights and promote collaborative problem-solving.

- Document and distribute lessons learned.
  Treat each AI project as an opportunity to codify what worked, what didn't, and why.

  Example: After each pilot, publish a short internal case study with key takeaways and recommendations.

- Apply the OODA loop to strategy execution.
  Use the OODA cycle—Observe, Orient, Decide, Act—to translate

learning into adaptive decisions and timely implementation. This approach keeps AI efforts aligned with evolving conditions while enabling structured iteration.

- Invest in ongoing AI readiness.
  Support learning through formal training, workshops, and external certifications.

  Example: Sponsor staff development on Gen AI or industry-specific ML tools.

Adaptive strategies are not just reactive—they are deliberately structured to evolve. By embedding continuous learning and applying the OODA Loop as a leadership discipline, organizations can scale AI initiatives with resilience, clarity, and sustained impact.

Case Study:
Leidos—Practical Application of the OODA Loop

This case study provides a high-level overview of how Leidos, a Fortune 500 federal contractor, successfully leveraged Gen AI to transform its IT support operations. It illustrates how large, complex organizations like Leidos can align AI initiatives with broader organizational goals to enhance productivity and maintain compliance in regulated environments. Additionally, the iterative principles of the OODA loop—Observe, Orient, Decide, Act—can be seen in the company's approach, demonstrating how this framework supports agile strategy and execution.

Company Background

- Headquarters: Reston, Virginia
- Industry: Technology Solutions Provider (Federal Contractor)
- Company Focus: Provides innovative solutions in national security, health, and other critical sectors.
- Customer Base: U.S. federal government and various global sectors.
- Workforce: 47,000 employees worldwide.

Challenges

Leidos faced significant challenges in managing IT support for 47,000 employees across mission-critical programs while operating in a highly regulated environment. These challenges included:

1. Managing Complexity at Scale: Supporting hundreds of enterprise applications while maintaining agility
2. Ensuring Federal Compliance: Adhering to strict security protocols to protect sensitive data
3. Improving Employee Productivity: Resolving routine IT issues quickly without overwhelming IT support teams

Brian Hobbs, VP of Enterprise Applications, described the task:

"We're not just focused on keeping things running, we aim to innovate and stay at the forefront, all while dealing with the immense scale and complexity of our operations."

Applying the OODA loop at Leidos

The strategic actions taken by Leidos closely mirror the OODA loop, providing a practical example of how this iterative framework drives success in complex environments.

1. Observe
   - Leidos identified key operational pain points by analyzing data on IT ticket volumes, employee satisfaction, and workflow inefficiencies.
   - It monitored advancements in AI-powered IT support solutions to evaluate its potential for addressing these challenges.
   - This phase allowed Leidos to gather the foundational insights necessary to frame its strategy.

2. Orient
   - The organization prioritized solutions that aligned with its goals of productivity, compliance, and scalability.
   - The company evaluated its technical infrastructure, existing platforms (Microsoft Teams and ServiceNow), and workforce needs to determine the best approach.
   - By analyzing trends and internal requirements, Leidos refined its focus to address routine IT issues through AI automation.

3. Decide
   - Leidos selected Moveworks's platform and implemented Iris, an AI-powered copilot, as the solution.
   - This decision balanced technical feasibility, compliance requirements, and the ability to scale efficiently.
   - Joe Cannon, Director of Intelligent Automation, emphasized:

     "We recognized the need for a solution that could scale with our operations, integrate seamlessly, and above all, keep us secure."

4. Act
   - Leidos deployed Iris to deflect routine IT tickets and automate workflows, enabling employees to resolve issues quickly without human intervention.
   - The company iterated the deployment based on early feedback, refining processes and expanding AI capabilities to achieve greater impact.

The Solution

Leidos's implementation of Iris focused on three key objectives:

1. Deflect Routine IT Issues: Use conversational AI to resolve common problems before they reach support teams.

2. Seamless Integration: Ensure the solution is integrated with existing platforms (Microsoft Teams, ServiceNow) for a frictionless user experience.

3. Compliance at Scale: Maintain strict adherence to federal security standards while rolling out the solution across the organization.

This approach allowed Leidos to enhance productivity while maintaining the necessary compliance protocols required in its industry.

Joe Cannon, Director of Intelligent Automation at Leidos, emphasized the company's strategy: "We recognized the need for a solution that could scale with our operations, integrate seamlessly, and above all, keep us secure. Moveworks provided that, allowing us to roll out AI quickly and efficiently."

Results

The deployment of Iris delivered significant benefits:

1. Reduced IT Workload: Routine IT issues were automatically resolved, freeing IT staff to focus on higher-value tasks.

2. Enhanced Employee Satisfaction: Employees resolved issues in real-time without disruptions, improving productivity and morale.

3. Maintained Compliance: The solution adhered to federal regulations, ensuring secure and compliant operations.

Hobbs remarked:

"Having a system that resolves issues before they escalate is a game-changer. It's like having a dedicated IT expert available 24/7."

Strategic Lessons

The success of Leidos highlights how organizations can use iterative frameworks like the OODA loop to navigate complexity and achieve their goals. Key takeaways include:

1. Aligning AI with Business Goals
   - Leidos implemented AI not for novelty but to solve specific challenges: improving productivity, reducing IT workloads, and maintaining compliance.
   - Takeaway: Identify strategic objectives before implementing AI solutions to ensure alignment and measurable impact.

2. Factoring in Compliance from the Outset
   - Operating in a regulated environment, Leidos prioritized compliance early, ensuring AI met stringent security standards without compromising functionality.
   - Takeaway: Security and compliance should be foundational elements of any AI strategy, especially in regulated industries.

3. Iterative Improvement for Productivity Gains
   - Leidos used feedback from early deployments to refine and scale its AI solution, continuously improving outcomes.
   - Takeaway: Apply iterative feedback loops to optimize AI performance and expand capabilities over time.

Unique Aspects of This Case Study

1. Federal Security Compliance
   Leidos deployed AI while maintaining compliance with federal regulations. This required a governed cloud architecture and secure handling of sensitive data.

2. Low-Code Customization for Agility
   The low-code nature of Moveworks's platform enabled Leidos to customize workflows rapidly and scale AI functionality across systems.

3. Proactive IT Issue Resolution
   Iris didn't just react to problems but proactively resolved them before they escalated, reducing the IT workload and enhancing employee productivity.

Encouraging Reflection:
How Can Your Organization Apply These Lessons?
While Leidos operates in a regulated environment, the lessons from its experience are broadly applicable. Consider the following questions:

1. Alignment with Goals: How can AI address pain points or improve efficiency in your organization?

2. Compliance and Security: Are there regulatory requirements to factor into your AI strategy?

3. Boosting Productivity: What routine tasks can AI automate to free up employees for higher-value work?

By observing your challenges, orienting AI solutions to your goals, making informed decisions, and acting iteratively, you can achieve similar success in your AI journey.

> Source: This case study is based on information provided by Moveworks (www.Moveworks.com)

Chapter Summary

Crafting an effective AI strategy requires a flexible, iterative approach that aligns with business objectives, adapts to evolving demands, and strengthens organizational capabilities. The guiding principles outlined in this chapter provide a structured framework for navigating AI adoption—ensuring strategic alignment, leadership engagement, and measurable outcomes.

The Leidos case study illustrates how iterative models like the OODA loop help organizations manage complexity, drive productivity, and stay focused on long-term value. By embedding continuous learning and adaptation into AI efforts, organizations position AI as a dynamic, scalable capability that evolves alongside their respective business.

Actionable Takeaway

Draft how each principle in this chapter applies to your organization's context. Identify one AI initiative that supports your business goals, and define success metrics to guide and measure its progress.

Next Chapter

Even the best strategies require disciplined execution. The next chapter will introduce a practical framework to guide your AI journey—from early experimentation to scaled impact.

Chapter 4

AI Maturity Model:
A Framework to Guide Your AI Journey

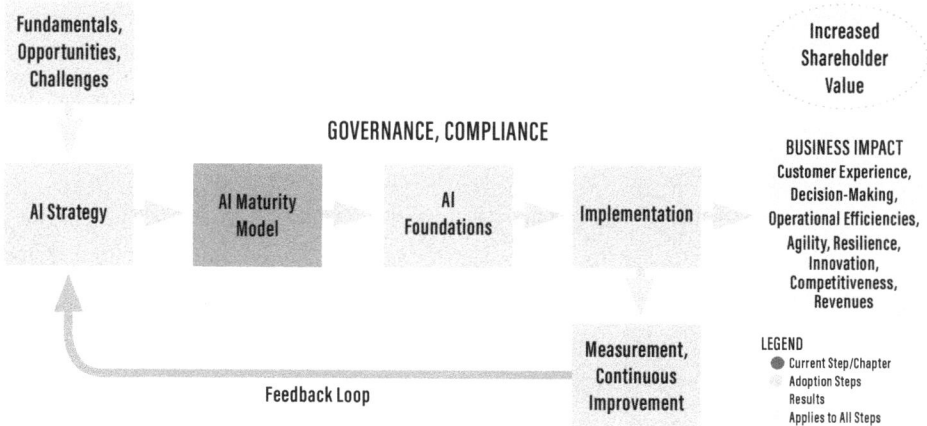

FIGURE 4.1 AI ADOPTION ROADMAP

uilding on the strategic foundations from the previous chapter, this chapter introduces the next stage of the AI Adoption Roadmap: the AI Maturity Model (see Figure 4.1). This framework helps assess your organization's current capabilities and define a path for progress. You'll explore each stage of the model, gain practical guidance for aligning AI with business goals, and learn from a case study featuring Palo Alto Networks and its measurable results.

The AI Maturity Levels

The AI Maturity Model outlines five distinct stages of AI adoption, serving as a roadmap for organizations to evolve from initial exploration to becoming industry leaders. Each stage emphasizes measured and scalable progress, ensuring that AI initiatives remain aligned with organizational objectives while advancing with maturity.

Level 1: AI Awareness (Exploration Stage)

At the AI Awareness stage, organizations begin exploring AI concepts and evaluating their potential impact. The focus is on building foundational knowledge, fostering AI literacy, and experimenting with small, low-risk projects that deliver tangible insights.

At this stage, organizations often seek approachable, low-barrier ways to experiment with AI. One of the most accessible and high-impact entry points is the use of LLMs, which provide immediate utility without requiring heavy technical investment.

Leaders should encourage their teams to move quickly to test, learn, and refine—continually identifying opportunities, executing pilots, and using these experiments to shape broader AI priorities. This ensures that learning remains practical, risks stay contained, and momentum is sustained.

LLMs—such as Open AI's ChatGPT, Microsoft Copilot, Google's Gemini, and Anthropic's Claude—enable organizations to rapidly explore AI's potential by automating routine queries, generating content, and delivering conversational interfaces. For example, deploying a simple chatbot powered by an LLM can effectively demonstrate AI's value, while also engaging leadership and stakeholders.

Key Actions

- Explain AI concepts using industry-relevant examples, while also addressing capabilities, potential risks, responsible use, and practical limitations to foster informed enthusiasm and healthy skepticism. This can be done through workshops, training sessions, awareness campaigns, informal Q&A sessions, or interactive demos.

- Introduce leadership and stakeholders to LLM tools to build familiarity and spark interest in AI, positioning them to champion further exploration.

- Launch simple LLM pilots in noncritical areas, such as drafting content or automating routine queries, to build confidence and demonstrate quick wins.

As organizations advance, they may wish to explore more sophisticated LLM applications or integrate these tools into operational workflows.

Level 2: AI Experimentation (Piloting Stage)

At this stage, organizations move beyond early exploration to conduct structured pilots that rigorously test where AI can deliver the most value. These pilots focus on specific, well-defined use cases aligned with strategic business objectives and may involve a range of techniques—from traditional ML and agentic systems to optimization models or hybrid approaches—chosen based on a problem rather than any single technology trend. The goal is to experiment within a controlled framework that manages risk, builds operational knowledge, and produces measurable outcomes to inform broader AI adoption.

Key Actions
- Establish an AI lab or dedicated innovation teams to provide a structured environment for experimentation across business functions. This creates a safe space to test ideas, explore both internal efficiencies and customer-facing opportunities, and build organizational capability for AI adoption.

- Implement small-scale pilots across departments such as customer service, marketing, and product R&D, prioritizing initiatives that address clear pain points or create direct customer value. Use these pilots to deliver tangible results, refine operating models, and inform your enterprise-wide AI strategy.

Level 3: AI Integration (Operational Stage)

AI becomes an integral part of daily operations in this stage, transitioning from isolated pilots to fully embedded workflows. Successful experiments are scaled to maximize impact across multiple business units, streamlining processes and driving measurable outcomes.

Key Actions

- Expand successful pilots into larger operations to achieve broader organizational benefits.

- Integrate AI with other technologies, such as predictive analytics and IoT, to strengthen decision-making.

- Develop governance frameworks to ensure ethical AI practices, emphasizing data privacy, bias mitigation, and transparency.

Level 4: AI Optimization (Strategic Stage)

At the Optimization stage, AI transforms into a strategic driver of innovation and differentiation. The focus shifts to enhancing AI systems for peak efficiency while ensuring alignment with long-term business objectives.

Key Actions
- Adopt advanced AI techniques, such as deep learning and NLP, to optimize performance.
- Regularly refine governance frameworks to address emerging ethical and regulatory considerations.
- Establish feedback loops to collect insights and continuously improve AI applications.

Level 5: AI Leadership (Transformative Stage)

At the AI Leadership stage, AI serves as a core enabler of organizational transformation and industry disruption. Organizations at this level leverage AI to redefine business models, drive innovation, and set industry benchmarks.

Key Actions
- Commit significant resources to AI research and development to pioneer new applications.
- Foster a culture of continuous learning and innovation, empowering teams to adapt and experiment. This involves approaches such as employee engagement, targeted training and upskilling, cross-functional collaboration, and cultivating a resilient, innovation-focused mindset. These strategies are presented in detail in Chapter 5 under "Change Management and Building an AI-Ready Culture," p. 111.

Stay ahead of industry trends by applying a disciplined, methodical approach to AI initiatives. This means continuously asking: What's emerging in the market? Is there a proven solution we can buy or adapt, or should we build it ourselves? Always scan the landscape first, as AI technologies evolve rapidly and external options may accelerate results. This approach aligns with the OODA loop introduced in Chapter 3—observe developments; orient them to your strategic context; decide whether to build, buy, or partner; and act through focused experimentation—ensuring your organization not only stays current, but also does so with deliberate strategic discipline.

AI Maturity Self-Assessment Checklist

A critical next step in the AI adoption journey is to accurately assess your organization's current position within the AI Maturity Model. Establishing this baseline enables the development of a strategic, actionable roadmap aligned with your capabilities and objectives. The Self-Assessment Checklist that follows is designed to support this process, helping you evaluate your maturity level and identify targeted priorities for advancement.

Level 1: AI Awareness (Exploration Stage)

1. AI Concept Familiarity
   - Have stakeholders been introduced to basic AI concepts relevant to your business?
   - Have workshops been conducted to build AI literacy among leadership?

2. Initial AI Projects
   - Have low-risk areas been identified for small-scale AI implementation?
   - Have Gen AI tools (e.g., chatbots, automated content creation) been piloted to demonstrate potential?

3. Curiosity and Engagement
   - Are team members exploring AI's potential impact?
   - Is an environment fostering curiosity about AI technologies in place?

Key Insight: Focus on foundational understanding while maintaining realistic expectations.

Level 2: AI Experimentation (Piloting Stage)

1. Pilot Project Implementation
   - Have pilot projects been launched in strategic departments (e.g., customer service, marketing)?
   - Are these pilots aligned with organizational goals?

2. Measuring Impact
   - Are metrics in place to evaluate pilot outcomes?
   - Have results been assessed to identify where AI can add value?

3. AI as a Strategic Tool
   - Is AI being used for both simple tasks and complex problem-solving?
   - Are plans in place to expand successful pilots?

Key Insight: Gather insights from pilot results to refine your AI strategy.

Level 3: AI Integration (Operational Stage)

1. Operational Integration
   - Have successful pilots been scaled and embedded into daily workflows?

- Are AI tools integrated into core business processes?

2. Governance and Ethics
   - Is there a governance framework for ethical AI use (e.g., data privacy, security)?
   - Are regular audits conducted to ensure compliance with governance policies?

3. Cross-Functional Collaboration
   - Are departments collaborating effectively to ensure seamless integration?
   - Are subject matter experts involved in refining AI applications?

Key Insight: Governance and collaboration are critical to scaling AI responsibly.

Level 4: AI Optimization (Strategic Stage)

1. Optimization of AI Systems
   - Are AI systems continuously refined to enhance performance?
   - Are advanced techniques (e.g., ML, NLP) being adopted?

2. Governance Refinement
   - Is the governance framework updated to address ethical and regulatory changes?
   - Is a culture of continuous improvement promoted within AI initiatives?

3. Feedback Loops
   - Are feedback mechanisms in place to refine AI based on real-world performance?
   - Is user feedback incorporated into AI system enhancements?

Key Insight: Optimization creates a strategic edge, making AI a core enabler of success.

Level 5: AI Leadership (Transformative Stage)
1. AI as a Core Business Driver
   - Has AI become central to your business model, driving innovation?
   - Are AI initiatives creating new opportunities and disrupting industry norms?

2. Investment in AI R&D
   - Is there substantial investment in AI research and development?
   - Are emerging technologies being integrated to keep pace with innovation?

3. AI-Centric Culture
   - Is there a culture of continuous learning and experimentation with AI?
   - Are teams empowered to lead AI-driven initiatives?

4. Industry Leadership
   - Are you recognized as a leader in AI adoption within your industry?

• Are you actively setting new standards for AI utilization?

Key Insight: At the leadership stage, competitive advantage comes from how an organization thinks and learns. The most successful organizations create cultures that reward curiosity and experimentation, and they embed disciplined practices—like the OODA loop—to consistently evaluate, adopt, and refine AI technologies. Leadership isn't about inventing AI; it's about being first to understand what matters, act decisively, and scale faster than the competition.

The self-assessment checklist is not just a starting point; it is a tool you can revisit regularly to track progress, evaluate AI initiatives, and identify opportunities for improvement. This iterative use ensures that your strategy and roadmap stay aligned with organizational goals, technological advancements, and market dynamics.

Case Study: Progression Through the AI Maturity Model—Palo Alto Networks

To demonstrate the practical application of the AI Maturity Model, this case study explores how Palo Alto Networks successfully navigated through its stages to achieve operational efficiency and strategic transformation.

Company Overview
• Headquarters: Santa Clara, California
• Industry: Cybersecurity
• Core Focus: Firewalls, cloud-native security tools, and solutions to combat cyberattacks

- Customer Base: About 85,000 enterprise customers across more than 150 countries
- Workforce: 15,000 employees worldwide

The Challenge

During a period of rapid growth, Palo Alto Networks doubled its workforce to approximately 15,000 employees, creating significant operational challenges. The surge in IT issues, HR requests, and policy inquiries became increasingly complex, especially with the onset of the COVID-19 pandemic. The transition to remote and hybrid work environments further amplified the need for seamless, 24/7 employee support.

Faced with these challenges, the company sought an innovative solution to scale its support operations effectively without compromising employee productivity or significantly expanding help desk staff.

The Solution

Palo Alto Networks implemented "Sheldon," an AI-powered bot developed by Moveworks, to automate employee support using NLP. Integrated across platforms like Slack, ServiceNow, and email, Sheldon enabled employees to resolve IT issues, HR inquiries, and policy questions via conversational AI.

This approach streamlined the support process by automating routine queries and tasks, allowing help desk staff to focus on more complex challenges. Employees gained immediate assistance whether working on-site or remotely, boosting productivity and overall satisfaction.

Results
- Scalability: Managed workforce growth without increasing help desk staff.
- Efficiency: Automated thousands of monthly employee requests, reducing response times.
- Productivity: Enabled help desk staff to focus on higher-value tasks by offloading routine queries to AI.

Progression Through the AI Maturity Model

Level 1: AI Awareness (Exploration Stage)

Palo Alto Networks began its AI journey by exploring potential applications for streamlining support operations. Leadership focused on building foundational AI literacy and assessing its value.

Actions
- Explored AI-driven solutions for IT and HR support.
- Evaluated the feasibility of automating routine queries with NLP-based tools.

Level 2: AI Experimentation (Piloting Stage)

The company piloted Sheldon for routine IT and HR inquiries. These small-scale trials validated AI's ability to reduce manual intervention and enhance support efficiency.

Actions

- Conducted pilot projects to automate basic support tasks.

- Focused on evaluating AI's effectiveness for routine, high-volume queries.

Level 3: AI Integration (Operational Stage)

After successful pilots, Sheldon was integrated into daily workflows, delivering 24/7 support across multiple platforms. AI became a key operational tool embedded in the organization's processes.

Actions

- Expanded Sheldon's deployment to platforms like Slack and ServiceNow.

- Fully integrated AI into workflows to provide real-time employee support.

Level 4: AI Optimization (Strategic Stage)

At this stage, Sheldon's capabilities were optimized to handle more complex HR and payroll-related tasks. The AI system became a strategic enabler, enhancing employee experiences and supporting rapid growth.

Actions

- Enhanced Sheldon to manage HR and payroll inquiries.
- Established feedback loops to continuously improve AI solutions.

Level 5: AI Leadership (Transformative Stage)

Palo Alto Networks positioned itself as an AI-driven leader by embedding AI into core operations. Sustained investment in AI allowed the company to maintain high productivity, even during growth and disruptions.

Actions

- Fostered a culture of innovation centered on AI solutions.
- Leveraged AI to sustain operational agility and employee satisfaction.

Case Study Lessons

- Strategic Alignment: Palo Alto Networks aligned AI initiatives with its goals, scaling support operations effectively during rapid growth.
- Iterative Progress: A phased approach through the AI Maturity Model enabled the company to test, refine, and optimize AI applications incrementally.
- Scalability and Innovation: AI enhanced support scalability while fostering innovation in workflows.

This case study exemplifies how organizations can transform operations and achieve industry leadership by strategically progressing through the AI Maturity Model.

How to Use the AI Maturity Model Strategically

As you move into the next stages of your AI journey, including building infrastructure, implementing solutions, establishing governance, and measuring success, keep the AI Maturity Model in mind as a reference point. The chapters ahead explore best practices and strategic decisions; however, how deeply you pursue them will depend on your organization's maturity. Use the model as a tool to calibrate your investments and priorities, ensuring each step aligns with your current capabilities and long-term vision.

Chapter Summary

The AI Maturity Model offers a clear, phased approach to adopting AI in a way that balances innovation with risk management. By progressing through five defined levels—Awareness, Experimentation, Integration, Optimization, and Leadership—organizations can scale AI initiatives thoughtfully, aligning them with strategic objectives and evolving business needs. This chapter broke down each stage with actionable steps and a self-assessment tool to help leaders evaluate their current state and plan next steps. The Palo Alto Networks case study illustrates how these stages come to life in a real-world setting, demonstrating how AI can evolve from a pilot concept into a transformative business capability.

Actionable Takeaway

Start by assessing your organization's AI maturity level using the self-assessment checklist. Identify where you currently stand and take strategic steps to progress toward the next stage. Regularly revisit the checklist to refine your AI strategy, optimize initiatives, and ensure continuous improvement in alignment with evolving business needs and technological advancements.

Next Chapter

A clear roadmap is essential, but it must be supported by readiness. In the next chapter, we will examine the technical, cultural, and organizational foundations needed to make AI implementation viable.

Chapter 5

Building the Foundations for AI

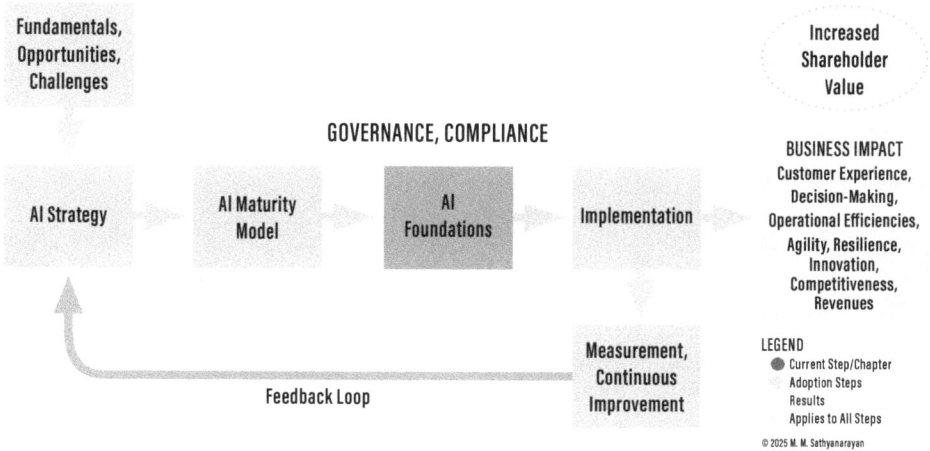

FIGURE 5.1 AI ADOPTION ROADMAP

With your AI strategy in place, focus areas prioritized, and organizational maturity assessed, this chapter advances to the next phase of the AI Adoption Roadmap: identifying the foundational capabilities required to execute at scale (see Fiigure 5.1). It presents a structured framework to assess readiness across people, platforms, and partnerships. This foundation enables the development of a practical roadmap to close capability gaps and support sustained AI delivery (see Figure 5.2).

AI Literacy

AI Staff

Technical Infrastructure

Data Management and Availability

Change Management, AI Ready Culture

External Partnerships

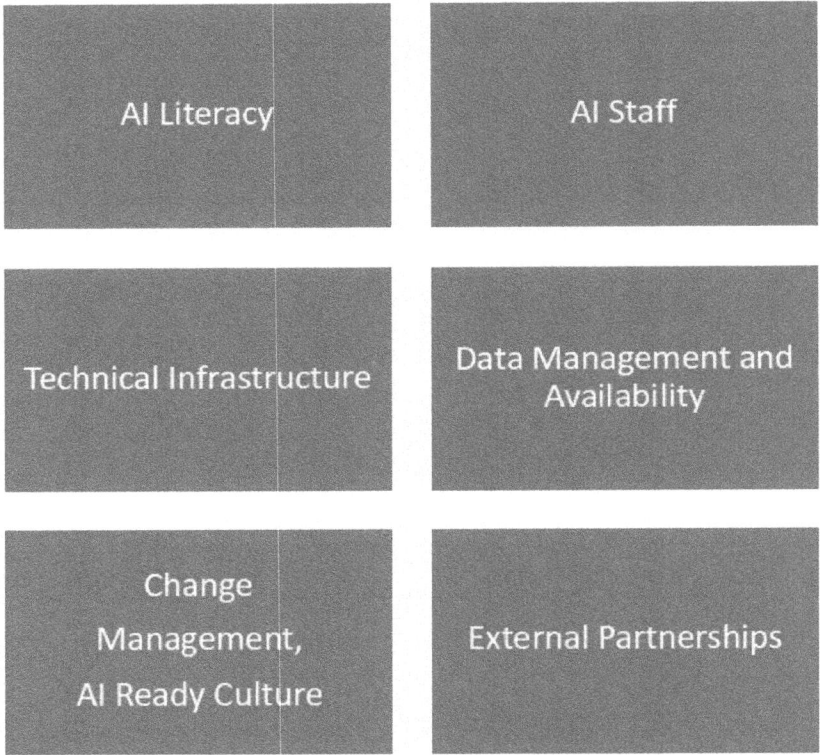

Figure 5.2 Foundation Areas for Scalable AI Execution

AI Literacy

As organizations advance their AI transformation journey, fostering a widespread understanding of AI across all levels becomes imperative. Here's how AI literacy can be enhanced:

- Comprehensive AI Education Programs: Implementing organization-wide education programs that clarify AI concepts and its technology, applications, and potential impact on different business aspects. Tailored training for various departments enhances understanding of how AI can specifically benefit their processes.

  At the same time, these programs should address the critical risks that come with AI adoption. As AI systems become more prevalent, so do threats like data leaks through careless prompt inputs, AI-generated phishing schemes, or misuse of sensitive organizational information. By incorporating topics on data privacy, cybersecurity implications, and how AI can be exploited by bad actors, AI literacy efforts help protect intellectual property, strengthen defenses against evolving threats, and ensure employees recognize their role in safeguarding the organization.

- Leadership in AI Understanding: Encouraging executives and managers to lead by example, promoting AI literacy through informed discussions and strategic thinking. This leadership helps set a positive tone for AI adoption throughout the organization.

- Cross-Departmental AI Workshops: Organizing workshops that bring together diverse teams to explore and discuss potential AI applications, enhancing cross-functional understanding and collaboration.

- Regular AI Updates and Case Studies: Sharing updates and case studies related to the organization's AI projects to provide insights, celebrate successes, and analyze challenges, fostering a culture of continuous learning and innovation.

- Encouraging Curiosity and Questions: Creating a supportive environment in which employees are encouraged to ask questions and express their thoughts on AI, facilitated through regular Q&A sessions and open forums.

- Real-World AI Problem-Solving Challenges: Engaging employees in hands-on AI problem-solving to make AI concepts more tangible and applicable to their work scenarios.

Critical Thinking as a Defense Against Blind Trust

As AI becomes embedded in day-to-day workflows, there's a significant risk that users will trust AI-generated outputs too readily—even when they are wrong. LLMs, for example, can produce hallucinated content that appears confident and coherent but may be factually incorrect. If users are not trained to critically evaluate these outputs, the organization risks making poor decisions based on flawed information. Therefore, AI literacy must go beyond tool usage to include structured training in critical thinking. Teaching employees how to question, verify, and contextualize AI outputs is essential to ensuring these tools support sound judgment rather than replace it.

AI Staff

Executing AI at scale requires a combination of business fluency and technical specialization. The following roles support implementation, oversight, and adoption across the enterprise.

Corporate AI Strategy Leadership
At the enterprise level, appoint a senior AI leader—such as a Chief AI Officer or Head of AI—responsible for defining and overseeing the organization's overall AI strategy. This role ensures AI initiatives are aligned with corporate objectives, ethical standards, and risk management practices, as outlined in Chapter 3. The AI leader serves as the central authority to coordinate efforts across business units, establish governance frameworks, and champion AI-driven transformation.

Product Owners within Business Units
Within each business segment, designate product owners who are embedded in their respective functions and accountable for translating corporate AI strategy into actionable, business-specific initiatives. These leaders identify and prioritize AI opportunities aligned with local objectives, work closely with functional teams to prepare for change, and ensure adoption delivers measurable value. Each product owner typically collaborates with one or more project managers who handle the detailed execution, timelines, and cross-functional coordination required to bring AI projects to life.

AI Project Manager

The AI Project Manager drives day-to-day execution of AI initiatives. This role oversees planning, coordinates cross-functional teams, and manages project timelines and milestones. The Project Manager serves as a bridge between technical contributors and business leaders—ensuring that delivery stays aligned with strategic priorities, risks are managed, and communications are clear.

Data Scientist

Data scientists are responsible for building and refining AI models. They analyze datasets, apply statistical and ML techniques, and collaborate with domain experts to solve business problems. They often iterate closely with business teams to ensure models are not only technically sound, but also usable and impactful.

ML Engineer

ML engineers take AI models to deployment. They focus on scalability, performance, and reliability—ensuring that models can run efficiently in production environments. They work closely with data scientists and IT to deploy, monitor, and optimize AI solutions.

Data Engineer

Data engineers design and maintain the data pipelines required to support AI models. They ensure data is accessible, clean, secure, and structured for analysis and training. Without a solid data foundation, AI initiatives stall—making this a foundational role.

AI Ethics and Compliance Advisor

This role ensures that AI development and deployment align with ethical standards, legal requirements, and organizational risk frameworks. Depending on the organization's structure, this function may sit within compliance, legal, or the AI governance team, but close coordination with technical and business teams is essential.

Domain Expert/Business Analyst

Domain experts ensure that AI systems are solving real problems. They provide context, test relevance, and help define success metrics. Business analysts often translate technical outputs into actionable business insights, helping ensure adoption and value creation.

Optional Roles Depending on Scope and Maturity

- AI UX Designer—Designs interfaces and experiences that make AI accessible and intuitive to end users.

- AI Trainer—Improves model behavior by fine-tuning or retraining using curated datasets and feedback.

- Prompt Engineer—Optimizes inputs to extract the best results from Gen AI models without retraining.

- MLOps Specialist—Supports the continuous integration and delivery of ML workflows, often in complex production environments.

- AI Educator—Builds AI literacy across the organization by designing and delivering training programs, workshops, and resources that equip employees at all levels with the knowledge needed to engage with AI tools effectively and responsibly.

These are general guidelines, and you need to set up the organization to suit your needs. For example, an AI educator role can be outsourced to begin with, and then you can staff up internally as your AI initiative matures.

Technical Infrastructure

A robust technical infrastructure is the backbone of any AI initiative. Without proper infrastructure in place, scaling AI effectively becomes a significant challenge. The following components are critical to ensure your infrastructure can support AI capabilities:

- Computing and Power Requirements
  AI models require significant computing and power resources to process large datasets, train ML models, and handle real-time applications. Evaluate your current infrastructure to determine if it meets the demands of AI.

- Data Storage and Management
  Your data storage solutions must not only accommodate large volumes, but also ensure fast, secure access. This includes implementing robust backup systems and data retrieval processes that allow AI systems to function efficiently.

- Cloud Computing Integration
  Cloud platforms such as AWS, Azure, or Google Cloud offer the flexibility to adjust resources as needed, making it an attractive option for organizations looking to scale their AI capabilities efficiently. For companies with limited on-premise infrastructure, cloud solutions can serve as a cost-effective alternative.

- Network Infrastructure
  AI models often process high volumes of data in real time. Your network should be capable of handling these data transfers without bottlenecks. Evaluate your current network performance and ensure it can support the high-bandwidth requirements of AI applications.

- Security and Privacy
  With AI relying heavily on sensitive data, ensuring cybersecurity and data privacy is critical. Your infrastructure should adhere to relevant data protection regulations and have strong cybersecurity protocols in place to safeguard AI systems and data.

- Cost Considerations
  Balancing infrastructure investments with long-term value is essential for sustainable AI deployment. Organizations must understand the distinction between capital expenditures (CapEx) and operational expenditures (OpEx) when planning their AI initiatives. Traditional on-premise infrastructure investments, such as purchasing servers and building data centers, typically fall under CapEx—incurring significant upfront costs but providing long-term assets. In contrast, cloud computing services are generally categorized as OpEx, allowing organizations to pay for computing resources on a flexible, consumption-based model. While OpEx can reduce upfront spending and increase scalability, it requires careful management to control long-term operating costs. Choosing the right balance between CapEx and OpEx models is critical to ensuring financial predictability, scalability, and alignment with your AI growth strategy.

Data Management and Availability

As emphasized earlier in this book, high-quality data is foundational to successful AI. This segment shifts focus from recognizing its importance to operationalizing it—outlining the concrete practices organizations must adopt to ensure their data infrastructure supports scalable, accurate, and ethical AI systems. Strong data management practices support both technical excellence and responsible AI deployment. Key principles include:

High-Quality and Reliable Data

AI systems require large, clean, well-labeled datasets. Ensuring data quality and integrity—accuracy, consistency, and protection from corruption—is essential for achieving reliable AI outcomes.

Bias Mitigation and Ethical Data Practices

AI models inherit biases present in training data. Regular auditing of datasets for potential biases (e.g., gender, race) is crucial to ensure fairness. Ethical practices must also be embedded, including compliance with data privacy regulations and transparency about how data is used, to build user trust and societal confidence.

Diversity and Coverage

Diverse datasets strengthen AI model robustness and generalization. Incorporating data from a broad range of sources helps models perform more accurately across different scenarios and populations.

Real-Time Processing and Scalability

Many AI applications, such as predictive analytics or dynamic decision-making, demand real-time data handling. Infrastructure must enable rapid data retrieval and high-speed processing to maintain AI performance as scale and complexity grow.

Change Management and Building an AI-Ready Culture

Successfully adopting AI requires more than technical execution—it demands a cultural shift that engages employees, aligns expectations, and prepares the organization for ongoing change. Traditional change management principles—such as transparent communication, leadership alignment, and continuous learning—still apply. However, AI introduces additional dynamics that make the change process more complex, including:

- The integration of opaque or evolving technologies into daily workflows
- The risk of real or perceived job displacement
- Shifts in team dynamics based on AI fluency
- The need for continuous upskilling and ethical awareness

The following section provides a consolidated approach to change management that integrates both conventional methods and AI-specific considerations so leaders can build trust, reduce resistance, and activate their workforce as a strategic asset in the transformation process.

Leadership Insight:
The Quiet Tension Around AI and Jobs

AI doesn't just introduce new tools—it changes workplace dynamics in ways many leaders underestimate. For the first time, knowledge-based roles—those centered on analysis, writing, decision-making, or client interaction—are directly affected by automation. These roles were largely untouched by previous waves of technology, but with AI, they are squarely in scope.

Even when roles are not eliminated, a new tension arises: the possibility of being outpaced or replaced by peers who adopt AI tools faster. This introduces a competitive pressure within teams that leaders must anticipate. The fear isn't always about layoffs—it's about relevance.

Over time, AI will likely create new jobs as past innovations have. But the short-term impact is uneven. Those in roles most immediately affected may be asked to help build or test the very tools that could redefine their future responsibilities. That's a psychologically complex situation, and one that calls for thoughtful leadership.

Change management must account for these human dynamics. Not just through training and communication, but through honest expectation-setting, targeted support, and meaningful involvement in the transformation process.

Leadership Commitment and Clear Communication

Leaders must go beyond setting strategic direction—they must visibly support the transition. This includes articulating the "why" behind AI initiatives, reinforcing the expected outcomes, and acknowledging uncertainty. Clear, consistent messaging helps teams understand how AI fits into the organization's broader transformation goals. Regular updates, two-way communication channels, and senior visibility build confidence and reduce resistance.

Employee Engagement, Training, and Upskilling

Developing an AI-ready workforce requires structured learning and role-specific support. Organizations should:

- Conduct skill gap assessments to determine AI readiness
- Offer tailored training covering both foundational concepts and job-specific applications
- Provide opportunities for mentorship, certification, and hands-on experimentation

When employees understand how AI affects their work—and how they can stay relevant—they are more likely to engage proactively.

Encouraging Cross-Functional Collaboration

AI initiatives benefit from diverse expertise. Cross-functional engagement helps surface blind spots, improve user adoption, and drive better business alignment. Leaders should:

- Facilitate joint pilots between technical and operational teams
- Host "AI use case" forums to explore real-world applications
- Encourage departments to co-develop solutions that reflect shared goals

This collaboration accelerates learning and fosters ownership across the organization.

Cultivating a Resilient, Innovative Culture
AI introduces new ways of working—and not every experiment will succeed. A resilient culture enables organizations to learn from failure, adapt quickly, and maintain momentum. Tactics include:

- Normalizing iteration and pilot testing as part of the process
- Celebrating milestones and highlighting early wins
- Supporting internal champions and recognizing contributions across functions

Resilience fuels long-term AI maturity.

Inclusive Participation and Employee Involvement
Trust and adoption grow when employees feel they are part of the change—not subject to it. Leaders should:

- Actively solicit input on AI implementation and experience
- Include frontline teams in testing and pilot refinement
- Communicate how feedback is influencing decisions

Inclusive processes increase transparency, improve outcomes, and reinforce psychological safety.

Embedding Ethics into the Change Process
Change management for AI must include a strong ethical foundation. As systems touch more sensitive or strategic processes, employees need confidence that guardrails are in place. Ethical integration means:

- Reinforcing fairness, transparency, and privacy standards
- Clearly assigning governance roles and escalation paths
- Addressing concerns about bias or misuse early and openly

Embedding ethics signals that AI is being implemented not just effectively—but responsibly.

External Partnerships: *Building AI-Specific Capabilities Through Strategic Collaboration*

External partnerships are a common lever in any transformation, but AI partnerships require a fundamentally different approach. Unlike traditional outsourcing or vendor arrangements, AI-related collaborations must navigate a fast-evolving technology landscape, ethical complexity, and deep technical integration.

To build long-term value and resilience through partnerships, organizations should focus on five distinct considerations:

- Specialized Expertise in a Rapidly Changing Field:
  AI partners often bring niche, cutting-edge strengths —
  such as foundation model tuning, generative AI solutions, or
  responsible AI auditing. These capabilities evolve at a fast pace,
  and many providers are still young or adjusting frequently to
  market shifts. Prioritize partners who combine depth in their
  specialty with the flexibility to adapt as the field matures.

- Transparency and Data Governance
  AI solutions often rely on proprietary models in which
  internal workings are opaque. Understand how data is handled, where models are hosted, and whether outputs can be
  explained and audited—especially in regulated industries.

- Co-Development and Knowledge Transfer
  Unlike traditional service delivery, AI partnerships should enable internal teams to learn and adapt. Prioritize collaborations that include joint pilots, co-design of models or prompts, and structured upskilling plans.

- Ethical Alignment
  Ensure partners share your standards for fairness, privacy, and responsible use. This is particularly important with Gen AI, where output can carry reputational and legal risks. Evaluate not just capabilities, but also values.

- Multimodal Ecosystem Management
  AI capabilities often come from multiple sources—cloud platforms, academic research labs, open-source contributors, and commercial providers. Each requires a different engagement model and risk profile. Map your partner ecosystem and assign clear accountability.

In short, AI partnerships are not just about cost or capacity; they are strategic enablers of learning, governance, and innovation. Organizations that approach partnerships with this lens will scale faster and more responsibly than those applying traditional outsourcing models.

AI Readiness Isn't About Checking Boxes

AI readiness assessments must go beyond standard IT or process maturity reviews. They require deeper reflection on how your organization learns, adapts, and governs technology that is both powerful and unpredictable.

The shift to AI introduces dimensions that traditional readiness models often overlook. These include the ethical complexity of model behavior, the opacity of outputs, the speed of evolution, and the organizational need to balance experimentation with accountability.

This section brings together those considerations into a single integrated lens—one that covers both conventional transformation readiness and the unique challenges of AI. Think of this not just as a checklist, but as a strategic mirror: it helps leadership teams identify friction points, uncover hidden gaps, and align around what it will truly take to scale AI with confidence.

Key Questions to Guide Your Readiness Assessment

- Do we have clearly defined ownership for AI across business, technology, data, and compliance teams?

- Are our data sources high-quality, well-governed, and ethically sourced?

- Can our infrastructure scale to support data-intensive and real-time AI applications?

- Have our workflows been redesigned to integrate AI outputs, not just bolt them on?

- Are employees equipped with the skills—and the confidence—to use AI tools effectively?

- Are we actively addressing bias, transparency, and explainability in our AI systems?

- Do our external partnerships support long-term capability-building, not just project delivery?

- Is there a feedback loop between pilot outcomes and strategic decision-making?

- Are risk management and innovation both represented in how we govern AI?
- Do we have executive alignment on the role of AI in our transformation priorities?

Use this list not only to evaluate your current state, but to guide leadership discussions about where to invest, how to sequence next steps, and what to treat as strategic nonnegotiables in your roadmap ahead.

Developing a Roadmap for AI Readiness

Once you've assessed your AI readiness and identified key capability gaps, the next step is to develop a roadmap tailored to the specific dynamics of AI initiatives. Unlike traditional IT programs, AI adoption is rarely linear. It evolves through cycles of experimentation, validation, and adaptation and is shaped by shifting data, emerging regulations, and iterative learning.

A successful AI roadmap reflects this reality. It must align tightly to business strategy while building in the feedback loops and resilience needed to absorb surprises and scale responsibly.

Building Your AI Roadmap

Prioritize Critical Gaps That Enable Success

All gaps are not created equal. Focus first on areas that directly impact AI viability, such as data quality, governance maturity, and executive alignment on ethical use. For example, if your data pipelines are unreliable or you lack clarity on who owns AI risk, those must be addressed before scaling. At the same time, ensure your roadmap strikes a balance between internal efficiencies and customer-focused innovation, creating new value that attracts customers and opens fresh revenue opportunities, not just streamlining internal processes.

Design for Iteration, Not Perfection

Many AI projects require cycles of trial and refinement before they can be deployed at scale. Build action plans that anticipate these phases, including checkpoints for model validation, user feedback, bias evaluation, and production readiness, not just go-live dates.

Set Milestones That Reflect Learning and Progress

AI milestones should measure more than whether a solution has launched. Include outcome-oriented checkpoints such as:

- Hitting accuracy thresholds

- Completing a fairness review

- Achieving a targeted reduction in cycle time

- Integrating stakeholder feedback into the next model version

These kinds of milestones track meaningful progress—even if the final model hasn't yet scaled.

Plan for Projects That Don't Scale—and Learn from Them

Every AI prototype will not make it to production. That's not failure; it's part of the process. Design your roadmap to treat early-stage pilots as structured learning efforts, with resources in place to capture insights and apply them to future initiatives.

Allocate the Right Specialized Roles

Ensure your roadmap is supported by people who understand AI's operational and ethical dimensions, which includes data scientists,

ML engineers, AI ethicists, prompt engineers, and support roles for ongoing tuning and retraining. Planning for post-launch care is just as important as development.

Make the Roadmap a Living Document

AI doesn't stand still, and neither should your roadmap. Build in regular checkpoints to review whether current projects still align with business priorities and emerging risks. Think of the roadmap not as a blueprint, but as a strategic compass anchored to long-term goals but responsive to changing conditions.

Chapter Summary

This chapter outlined what it takes to move from strategy to execution by building the foundational elements needed for AI adoption at scale. These include workforce readiness, cross-functional collaboration, infrastructure fit for purpose, and a roadmap that reflects both strategic priorities and AI's unique demands. Effective transformation isn't driven by intent alone; it also depends on cultural alignment, operational readiness, and the capacity to learn and adapt.

Actionable Takeaway

Conduct a focused audit of your organization's readiness across talent, infrastructure, and workflows. Identify capability gaps that could slow AI execution, and build a roadmap that prioritizes what matters most, starting with AI literacy and trustworthy data pipelines.

Next Chapter

With your foundation in place, the next step is action. The following chapter will focus on execution—how to launch pilots, deliver early wins, and create momentum as you move toward enterprise-wide AI adoption.

Chapter 6

AI Implementation: Turning Strategy into Reality

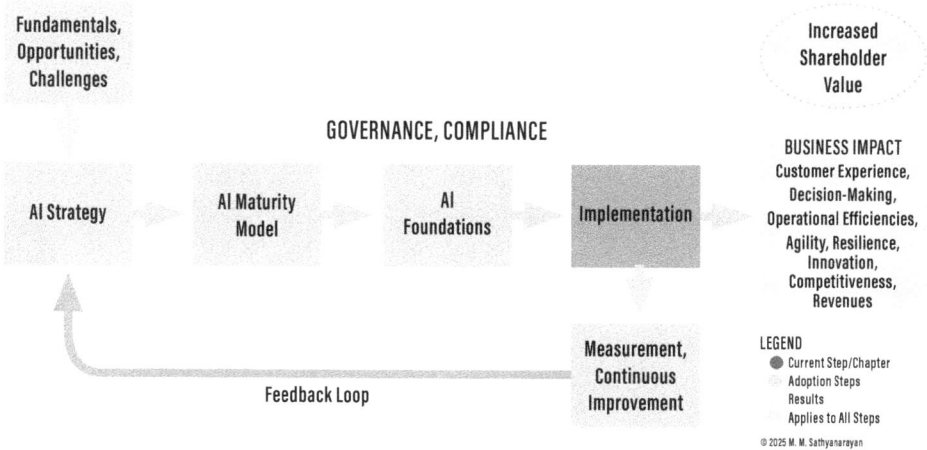

Fundamentals, Opportunities, Challenges

Increased Shareholder Value

GOVERNANCE, COMPLIANCE

AI Strategy → AI Maturity Model → AI Foundations → Implementation

BUSINESS IMPACT
Customer Experience,
Decision-Making,
Operational Efficiencies,
Agility, Resilience,
Innovation,
Competitiveness,
Revenues

Measurement, Continuous Improvement

Feedback Loop

LEGEND
● Current Step/Chapter
Adoption Steps
Results
Applies to All Steps

© 2025 M. M. Sathyanarayan

FIGURE 6.1 AI ADOPTION ROADMAP

Successfully adopting AI requires translating strategic goals into actionable steps. This chapter moves us into the implementation stage of the AI Adoption Roadmap, in which plans become practice (see Figure 6.1). It provides a practical guide for putting AI into action, focusing on five key areas:

1. Streamlining Processes Before Applying AI

2. Selecting the Right AI Projects

3. Key Steps for Successful AI Pilots

4. Managing AI Projects with a Portfolio Mindset

5. Strategic Choices for AI Tools and Technologies

1. Streamlining Processes Before Applying AI

Because AI systems are probabilistic rather than rule-based, their failures are harder to trace and correct. Unlike traditional software, which throws clear errors when rules break, AI may silently produce flawed outputs—often without any obvious indication that something is wrong.

These flaws typically emerge from the data and workflows AI learns from. If the underlying process is inconsistent, biased, or poorly defined, those issues can be amplified at scale. Worse, they may go unnoticed until performance degrades or harm is done, making flawed processes a hidden risk multiplier.

That's why streamlining isn't a best practice; it's a prerequisite. Before introducing AI into a workflow, ensure it is logically sound, consistently followed, and well understood. Clean, well-governed processes don't just make AI work better; they reduce the risk of unintended consequences and enable more reliable outcomes.

2. Selecting the Right AI Projects

Know when AI Is the Right Tool

At this stage in your AI journey, the question is not: "Can we use AI?" but "Should we use AI here?" Project selection is no longer about comparing AI to traditional systems; it's about identifying use cases in which learning, pattern recognition, or language understanding create a strategic advantage.

Focus your AI efforts where:

- Data is abundant and high quality, enabling meaningful learning.

- Outcomes improve over time with iteration, feedback, or personalization.
- Human insight is valuable but limited in scale, such as real-time decision-making or language-driven tasks.

Avoid deploying AI in areas with:

- Unstable or highly ambiguous data environments in which feedback loops are weak.
- Well-defined processes in which rule-based systems can solve the problem faster and cheaper.

In short, the right AI project isn't just technically feasible; it's strategically valuable and organizationally ready.

Insights from Early AI Adopters

To help you select the right projects, consider insights from successful applications across industries. A survey of corporate directors revealed early gains by applying AI in the following areas:

- 33%: Data and reporting capabilities
- 22%: Customer service and support
- 19%: Workforce productivity
- 14%: Marketing and communications
- 12%: Product/service enhancements

These early wins highlight AI's ability to drive efficiency and productivity when applied to focused, well-scoped problems.

The survey also points out that AI is not yet delivering transformative impact at scale, largely due to the challenges of organizational readiness, data infrastructure, and tailored application development.

Looking beyond these early implementations, directors surveyed identified three key areas in which they see the greatest potential for AI to deliver significant value in the future:

1. Enhancing Products and Services: Many directors expect AI to drive innovation by creating new service offerings, automating complex processes, and improving customer experiences. This is particularly impactful in service-driven industries like financial services and healthcare.

2. Supporting Marketing and Communications: AI's ability to create high-quality, personalized content at scale positions it as a valuable tool for marketing teams. Early adopters report improved customer engagement through targeted campaigns and reduced operational costs via automation.

3. Driving Workforce Productivity: Directors foresee AI automating more sophisticated workflows, enabling employees to focus on strategic problem-solving and innovation. Achieving this potential will require sustained investment in training and change management to integrate AI into broader business processes effectively.

By understanding the distinction between early gains and future opportunities, organizations can prioritize projects that align with their immediate capabilities while building toward transformative applications.

Source: Corporate Board Member and Diligent Institute

Risk Level as a Selection Criteria

To select the right AI projects you need to consider risk, readiness, and the organization's ability to manage uncertainty. Many organizations use a simple framework to group projects into three risk tiers, helping align project choices with governance maturity and delivery capabilities.

This isn't a strict classification system; it's a planning aid. Use it to assess where to start, what to pilot, and how fast to scale.

1. High-Risk, High-Impact Applications (Human Collaboration-Critical)

   Examples: Medical diagnostics, legal advice, hiring decisions

   These require careful oversight due to ethical, legal, or reputational stakes. Start small, test under human supervision, and prioritize transparency and explainability.

2. Moderate-Risk, Autonomous Applications (Human-Supported Autonomy)

   Examples: Autonomous vehicles, financial forecasting

   These operate independently but still benefit from human fallback and performance monitoring. Pilots should include contingency planning and escalation protocols.

3. Low-Risk, Autonomous Applications (Autonomous Efficiency)

   Examples: Product recommendations, demand forecasting, email sorting

   These are ideal candidates for early adoption and fast iteration. They offer quick wins with manageable downside, helping teams build experience and confidence.

3. Key Steps for Successful AI Pilots

A tightly scoped pilot lets teams observe how AI systems behave under real-world conditions, validate business value, and uncover potential issues early. From there, a structured, phased implementation allows organizations to scale what works—step by step—while actively managing risks and adapting along the way.

The following checklist offers a practical approach to designing and executing AI pilots that build confidence, surface risks early, and prepare your organization for scalable deployment.

1. Define Success Criteria Clearly
    - Set measurable, objective targets aligned with business goals.
    - Example: In a customer support chatbot pilot, success could be:
        - 20% reduction in average response time
        - First-contact resolution rate increase from 65% to 80%
        - Customer satisfaction scores exceeding 4.5 out of 5 within three months
    - *Note:* In AI pilots, success criteria may also include model performance metrics like precision, recall, or confidence intervals, not just business KPIs.

2. Select the Right Pilot Project
    - Choose a project that is manageable in scope but material enough to demonstrate meaningful impact.
    - Look for processes in which:
        - Variability or complexity benefits from AI
        - Data is available and of reasonable quality
        - Outcomes can be measured objectively

3. Build a Cross-Functional Team
   - Assemble AI and data science experts, domain specialists, business stakeholders, and change management leaders.
   - AI projects succeed when technical feasibility, business needs, and human adoption are addressed together.

4. Prepare High-Quality, Relevant Data
   - Ensure data is clean, complete, and meets privacy and compliance standards.
   - Pilot data should reflect realistic use cases, not cherry-picked ideal conditions to give the model meaningful learning opportunities.

5. Scale in Phases, Not All at Once
   - Expand successful pilots incrementally, incorporating real-world feedback to refine both model and operational processes.
   - Early production use may reveal model drift, bias emergence, or unexpected edge cases that were not evident during pilot training.

6. Engage Stakeholders and Manage Risks Proactively
   - Communicate frequently and transparently with internal stakeholders.
   - Manage expectations around what AI can realistically achieve.
   - Actively monitor new risks, especially around data integrity, bias, and ethical impacts as the model moves closer to full deployment.

4. Managing AI Projects with a Portfolio Mindset

Managing AI initiatives with a portfolio mindset helps balance innovation, risk, and resource allocation across projects at different maturity levels. Since AI projects carry higher level of risk, a portfolio approach allows leaders to hedge against that uncertainty while still advancing business goals.

Key Principles
1. Balance Risk Across Projects
   - Balance low-risk, high-confidence use cases (e.g., recommendation engines) with higher-risk, exploratory projects (e.g., predictive modeling for new markets).
   - Include both production-ready AI and experimental pilots to maintain forward momentum without overexposing the organization.

2. Treat Failures as Learning Assets
   - Use failed pilots or underperforming models as opportunities to improve data readiness, fine-tune methods, or use case targeting.
   - Establish post-mortem reviews as a routine part of the AI development cycle.

3. Prioritize Investment Based on Readiness
   - Invest in projects that are not only high-impact but also supported by available, high-quality data and business engagement.
   - Avoid overcommitting to projects that depend on speculative data, unclear outcomes, or unproven vendors.

4. Encourage Experimentation and Learning
   - Encourage cross-functional teams to test new AI techniques and explore unconventional data sources.
   - Frame AI work as iterative—learning over time is part of the process, not a sign of failure.

Example

A financial institution applied this mindset by balancing quick-win projects like customer service chatbots with long-term investments in AI-driven fraud detection. While chatbots delivered immediate ROI and improved customer responsiveness, the fraud detection initiative required deeper data work, regulatory alignment, and iterative model development—ultimately delivering strategic value over time. This mix allowed the organization to realize early wins while de-risking longer-term innovation.

Treat your AI initiatives like a portfolio—balancing short-term gains with long-term capability building, and tactical pilots with strategic bets. Not all models will succeed, but a disciplined, diversified approach ensures continuous learning and sustained value.

5. Strategic Choices for AI Tools and Technologies

Selecting the right AI tools and vendors is critical to successful implementation. These decisions shape the scalability, adaptability, and long-term impact of your AI initiatives.

Your organization's AI maturity level should guide these choices. Early-stage adopters may require flexible, low-code platforms with vendor support, while more mature organizations may prioritize tools that enable deeper customization, control, and integration.

Key Considerations for AI Tools and Vendors

Before deciding whether to build or customize an AI solution, start by surveying the market. The pace of AI innovation has led to an explosion of tools, platforms, and niche applications that may already solve your business challenge—or at least serve as a starting point. Avoid reinventing the wheel unless doing so creates meaningful differentiation or addresses a gap no existing solution can fill.

1. Customization versus Ready-Made Solutions
   - Custom AI Development: Enables organizations to build models tailored to specific business problems, data environments, and workflows. While this allows for greater precision and differentiation, it also demands more resources, specialized talent, and longer lead times with added risks due to model training variability.
   - Out-of-the-Box AI Tools: These are prebuilt solutions designed for common tasks like classification, forecasting, or summarization. They offer fast deployment and lower entry costs, but may lack the flexibility to adapt to domain-specific data or evolving business needs, often requiring workarounds or limited customization.
   - Pre-Trained Models: Offer a balance between speed and adaptability. Models such as LLMs or vision transformers can be fine-tuned with organization-specific data, shortening time to value while maintaining relevance to your unique context.

     Think of this as similar to onboarding a new college graduate: the model brings general capabilities but only becomes effective after exposure to your business environment. Just as new hires must learn workflows, develop role-specific skills, and grow through experience, pre-trained models require fine-tuning, targeted training, and ongoing updates to perform reliably and stay relevant.

2. Data Readiness and Vendor Support
   Even strong AI models falter without reliable, well-structured data. Evaluate vendors not only on model quality, but on how well they help you operationalize data.
   • Vendor-Provided Data: Some vendors offer baseline datasets or pre-configured tools that accelerate deployment. This can be helpful when internal data is limited or not immediately usable.
   • Hybrid Approaches: Seek vendors that allow you to enhance their models using your own business-specific data for greater accuracy and relevance.
   • Data Governance and Security: Ensure vendors meet your standards for data privacy, integrity, and regulatory compliance, particularly in sensitive domains.

3. Transparency and Accountability
   Avoid opaque systems. Choose tools that offer visibility into how models work, how decisions are made, and how performance is measured—especially when compliance, trust, or explainability are essential.

4. Domain Expertise and Track Record
   Vendors with experience applying AI in your industry are better equipped to navigate domain-specific data challenges, regulatory nuances, and operational complexity. Their models are more likely to reflect real-world patterns relevant to your business and avoid pitfalls such as biased

training data, noncompliant features, or unrealistic performance assumptions.

5. Long-Term Flexibility and Scalability
   AI solutions must evolve alongside your organization—
   not just in terms of user volume, but also in adapting to
   new data sources, use cases, and regulatory expectations.
   Look for platforms that support ongoing model retraining,
   cross-functional deployment, and seamless integration into
   a growing AI ecosystem rather than tools that lock you into
   narrow, static use cases.

Financial Considerations Across Maturity Stages

- Early adopters benefit from cloud-based, pay-per-use tools
  that reduce upfront investment and enable experimentation.

- Scaling organizations should evaluate whether to shift toward
  capital investments (e.g., on-premise infrastructure) or maintain an operational expenditure model through cloud services.

- Mature AI users often adopt hybrid models—balancing control, performance, cost efficiency, and compliance needs.

**Vendor Selection: Beyond Technology—Support, Transparency
and Ethics**

As AI becomes embedded in business-critical processes, the stakes
rise. Vendor selection must reflect more than technical fit—it must
account for trust, ethics, and long-term viability.

Key factors to assess include:

- Ethical transparency—Are decisions explainable? Is data
  usage disclosed and compliant with your ethical standards?

- Bias and fairness—Has the vendor conducted bias testing? Can the system flag or mitigate inequitable outcomes?

- Regulatory alignment—Is the vendor compliant with applicable laws such as GDPR, CCPA, HIPAA, or sector-specific regulations?

- Vendor accountability—Are models clearly documented? Are performance metrics transparent? Are service-level agreements (SLAs) well-defined?

- Industry expertise—Has the vendor demonstrated success in contexts similar to yours?

- Ongoing innovation—Does the vendor maintain regular updates and demonstrate commitment to R&D?

Building Balanced Vendor Relationships

Vendors play a key role in accelerating adoption, but overreliance can introduce risk. Strong internal literacy allows organizations to evaluate offerings critically, maintain control over outcomes, and ensure that vendor engagements remain strategic rather than dependent.

Sustainable AI adoption depends not just on the tools you choose, but on the partnerships you build—and your ability to use them with confidence and control.

Chapter Summary

This chapter outlined how to move from AI strategy to action. It covered key execution priorities: streamlining business processes to avoid automating inefficiencies, selecting AI projects aligned with business goals, piloting AI in controlled environments, managing implementation risks with a portfolio mindset, and choosing tools and vendors that align with your AI maturity. The central message: successful AI adoption is not about big launches; it's about disciplined, iterative execution that delivers measurable value while building long-term capability.

Actionable Takeaway

Choose one business domain and run a tightly scoped AI pilot. Focus on streamlining the underlying process first, then apply AI to create measurable improvements. Use the experience to validate your implementation model and build momentum for broader rollout.

Next Chapter

Execution is only part of the equation—what ensures trust, safety, and long-term viability is how you govern AI. In the next chapter, we will elevate governance from a compliance afterthought to a strategic, risk-based discipline—one that demands attention from senior leadership and adapts as your AI capabilities evolve.

Chapter 7

AI Governance and Legal and Ethical Compliance

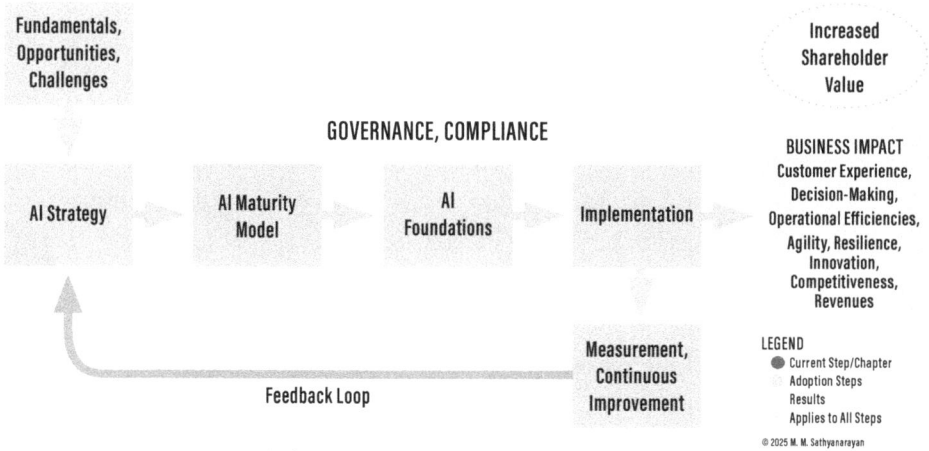

FIGURE 7.1 AI ADOPTION ROADMAP

Governance as a Cross-Cutting Risk Management Discipline

With your AI foundations in place and pilots underway, one critical discipline must now operate in parallel: governance. More than a compliance function, AI governance is a form of strategic risk management that ensures trust, reduces liability, and creates the conditions for safe, scalable innovation.

As shown in the AI Adoption Roadmap (see Figure 7.1), governance is not a one-time step—it is a cross-cutting capability that must evolve alongside AI maturity. Traditional compliance frameworks, built for static rule-based systems, fall short when applied to adaptive, probabilistic models that evolve in real time.

AI failures are not hypothetical. Real-world examples include models that denied qualified candidates job opportunities due to biased training data, issued racially skewed credit decisions, or produced misleading outputs—all without human oversight. These risks are amplified by AI's scale, speed, and opacity.

This chapter introduces a flexible and risk-aligned approach to governance called Minimum Viable Governance (MVG). MVG enables organizations to start with essential safeguards and scale oversight as maturity—and exposure—increases. It aligns directly with the AI Maturity Model presented earlier, ensuring governance grows in proportion to your AI vision.

MVG: *A Risk-Aligned Solution*

To address this challenge, organizations need a governance model that can scale with AI adoption and evolve with its risks. MVG provides that foundation. Inspired by Agile and Minimum Viable Product (MVP) principles, MVG emphasizes starting with core governance controls, then expanding oversight iteratively based on actual risk and usage—not hypothetical worst-case scenarios.

Rather than creating rigid, top-down governance structures from day one, MVG encourages organizations to:

- Prioritize governance in high-risk use cases (e.g., hiring, lending, healthcare)

- Begin with lightweight, essential controls—such as documentation, accountability, and risk assessment

- Scale governance alongside AI maturity, complexity, and impact

In essence, MVG reframes governance not as a compliance burden, but as a flexible, risk-managed enabler of responsible AI adoption.

Aligning MVG to AI Maturity Levels

MVG is designed to align directly with an organization's AI maturity level—a concept introduced in Chapter 4. Table 7.1 outlines how governance priorities evolve alongside AI maturity:

Table 7.1 Governance Priorities and AI Maturity

AI Maturity Level	Governance Focus	Key Actions
Level 1: AI Awareness	Basic AI Ethics & Data Oversight	Establish AI principles; conduct risk-awareness training
Level 2: AI Experimentation	Risk-Tiered AI Governance	Assign governance ownership; introduce basic risk assessments
Level 3: AI Integration	Operationalized Governance	Implement bias detection, model documentation, and explainability
Level 4: AI Optimization	Compliance & Audit Readiness	Align with EU AI Act, NIST AI RMF and others as applicable; implement regular audits
Level 5: AI Leadership	Enterprise-Wide Governance Strategy	Deploy real-time risk dashboards; lead in industry governance

Key AI Governance Risks and How MVG Addresses Them

The following are the most critical categories of AI-related risk and include how MVG enables organizations to respond in a proportionate, scalable way.

1. Data Governance Risks
Poor data governance leads to skewed or discriminatory outcomes, privacy violations, and legal noncompliance. In regulated industries, failure to control data lineage or access can trigger serious penalties under laws like GDPR and CCPA.

MVG starts with basic data controls—such as data source documentation, access restrictions, and consent tracking. As maturity increases, organizations can layer in more advanced capabilities such as data provenance auditing, role-based data access, and automated compliance checks.

2. Bias and Fairness Risks
AI can unintentionally amplify biases present in training data, resulting in unfair or discriminatory outcomes. This has already led to high-profile cases in hiring, lending, and law enforcement—damaging trust and triggering regulatory action.

MVG encourages organizations to prioritize fairness audits in high-risk use cases early on. Lightweight bias testing and fairness documentation are introduced in early maturity stages. Over time, these evolve into more formalized governance tools—such as bias detection pipelines, fairness KPIs, and external audits.

3. Transparency and Explainability Risks

Many AI models, especially deep learning systems, lack transparency. When decisions cannot be explained to regulators, customers, or internal team, trust erodes.

MVG supports graduated transparency. Early stages focus on model documentation and basic decision logging. As systems scale, explainability becomes more rigorous with model cards, interpretable-by-design architectures, and compliance with explainability mandates.

4. System Evolution and Model Drift

AI models degrade over time as real-world data shifts. This phenomenon, known as "concept drift," can cause performance to decline or lead to unintended behaviors—without triggering alerts.

MVG introduces model performance monitoring early—tracking drift indicators and confidence levels. In later stages, this evolves into continuous validation, automated retraining governance, and real-time dashboards.

5. Regulatory Noncompliance

AI regulation is rapidly evolving. Noncompliance with laws such as the EU AI Act, NIST AI RMF, or sector-specific mandates can lead to fines, operational bans, and reputational fallout.

MVG ensures baseline regulatory alignment even at low maturity levels. As adoption scales, organizations can build in formal compliance reviews, audit preparedness, and external certifications.

6. Shadow AI and Unauthorized Use

AI developed outside formal governance channels can introduce compliance blind spots, security issues, or legal exposure. This "Shadow AI" may be invisible to corporate oversight until a problem arises.

MVG supports opt-in, lightweight mechanisms to bring Shadow AI under control, starting with inventory tracking and self-reporting and progressing to tiered risk-based oversight.

7. Third-Party AI and Supply Chain Risk

Relying on vendors does not transfer risk. Many external AI models lack transparency, increasing exposure to bias, compliance failures, or cybersecurity vulnerabilities.

MVG encourages organizations to require risk certifications, conduct vendor audits, and introduce model-level accountability in supplier contracts.

8. Gen AI and Content Authenticity

Gen AI tools create new governance issues around misinformation, copyright, and content traceability. Without safeguards, they can damage trust and violate legal standards.

MVG enables organizations to roll out policies for usage, watermarking, verification workflows, and human oversight in a phased, scalable way.

Governance Ownership and Accountability Under MVG

AI governance only works when ownership is clear. Yet many organizations struggle with deciding whether it belongs to IT and Data Science, Legal and Compliance, or Business Leadership.

MVG supports a distributed ownership model that evolves with AI maturity:

- Legal and Compliance: Oversee policy, ethics, regulatory alignment
- IT and Data Science: Own technical integrity, security, fairness
- Business Leadership: Aligns AI efforts with strategy and accountability
- Cross-Functional Teams: Coordinate enterprise-wide governance in later stages

At early maturity levels, governance may be handled by a small team. As complexity grows, so should ownership structures—without overburdening early adopters.

The Path Forward: *Holistic AI Governance*

As AI becomes embedded in more business processes, governance must scale with it—not just to meet compliance requirements, but to safeguard brand, trust, and long-term impact.

MVG provides a practical path to doing just that. It helps organizations govern AI:

- Proportionally to risk
- Aligned with maturity
- Without blocking innovation

By embedding MVG into your AI strategy, you create a governance model that grows with you—supporting ethical, effective, and scalable adoption. Now we will look at a real-world case study.

Case Study: *Avanade's Approach to AI Governance*

To see these governance principles in action, we examine how Avanade, a leading digital innovator with over 60,000 employees implemented a scalable, risk-aligned AI governance model. Its approach demonstrates how organizations can successfully balance compliance, transparency, and workforce enablement using MVG principles.

Centralized Oversight: The Control Tower

Avanade implemented a centralized system, referred to as the "Control Tower," to monitor AI integration and associated risks across various organizational levels. This approach ensures alignment between board-level governance directives and operational practices, facilitating comprehensive oversight and accountability.

Comprehensive Training: The AI Driver's License

Recognizing the importance of role-specific training, Avanade developed the "AI Driver's License" program to equip employees with the necessary knowledge for responsible AI usage. The program includes:

- Foundational Training: All employees undergo basic training on responsible AI use, covering essential principles and organizational policies.

- Advanced Modules: Specialized training modules are designed for specific roles; for instance, data scientists receive in-depth instruction on compliance and ethical considerations pertinent to their work.

- Periodic Assessments: Licenses are subject to regular evaluations to ensure continuous learning and adherence to evolving best practices.

Motivating Employees: AI as an Enabler

To foster a culture of innovation and trust, Avanade positions AI as a tool for personal and professional growth. Strategies to motivate employees include:

- Showcasing Success Stories: Highlighting individual and team achievements in AI-driven projects to inspire and encourage broader participation

- Incentivizing Innovation: Offering rewards and recognition for innovative applications of AI that align with organizational goals and ethical standards

Avanade's practices demonstrate how a centralized and well-structured approach to AI governance can align board-level priorities with day-to-day operations, fostering accountability and trust at scale.

Chapter Summary

Effective AI governance is not a static checklist; it must evolve alongside an organization's AI maturity. As regulatory scrutiny intensifies and AI technologies grow more complex, governance must expand from foundational policies to advanced, real-time oversight. The MVG model offers a practical framework for scaling governance efforts in a way that is proportional to risk and aligns with business strategy. From addressing emerging challenges like Shadow AI and generative models to establishing clear accountability structures, the chapter emphasized that governance is both a risk management discipline and an enabler of responsible innovation.

Actionable Takeaway

Conduct an AI Governance Readiness Assessment to identify current risk and compliance gaps, then apply the MVG model to build a scalable, maturity-aligned governance framework that integrates seamlessly into your AI strategy.

Next Chapter

Having addressed the responsibilities that come with AI, the next challenge is determining whether your efforts are producing meaningful results. The following chapter will introduce the critical question of measurement—and what success truly looks like in an AI-enabled organization.

Acknowledgment

This chapter draws inspiration from OpenAI's principles and insights from the 2024 Boardroom Summit featuring Florin Rotar, Chief AI Officer at Avanade. These perspectives have been adapted to provide actionable strategies for enterprise managers navigating AI governance.

Chapter 8

Measuring AI Success

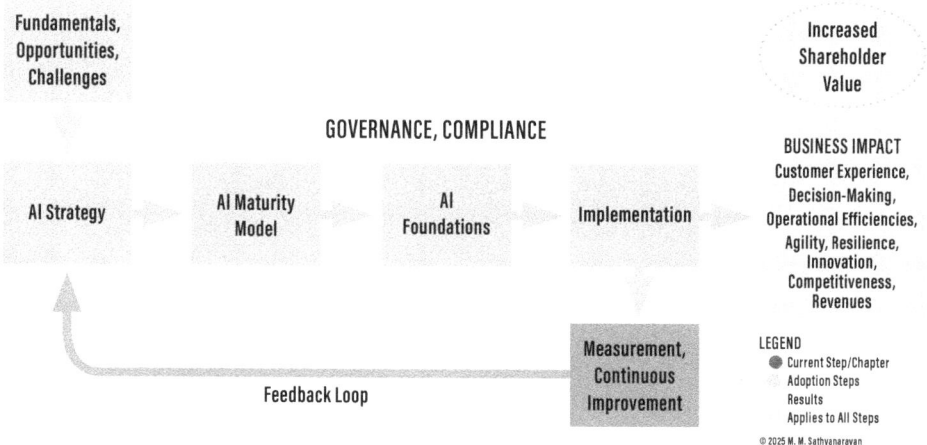

Fundamentals, Opportunities, Challenges

GOVERNANCE, COMPLIANCE

Increased Shareholder Value

AI Strategy

AI Maturity Model

AI Foundations

Implementation

BUSINESS IMPACT
Customer Experience, Decision-Making, Operational Efficiencies, Agility, Resilience, Innovation, Competitiveness, Revenues

Measurement, Continuous Improvement

Feedback Loop

LEGEND
● Current Step/Chapter
Adoption Steps
Results
Applies to All Steps
© 2025 M. M. Sathyanarayan

FIGURE 8.1 AI ADOPTION ROADMAP

easuring the success of AI initiatives requires an approach that reflects their evolving and transformative nature. Unlike static technologies, AI grows with an organization, shifting its value proposition as capabilities mature.

This chapter moves into the measurement and improvement stage of the AI Adoption Roadmap, in which the focus shifts from experimentation to evaluating performance and impact (see Figure 8.1). In early adoption, metrics emphasize building foundational capabilities: fostering expertise, embedding AI in workflows, and nurturing an innovation-ready culture. As maturity increases, measurement pivots toward tangible business outcomes like operational efficiency, decision quality, and market advantage.

This progression reflects key principles introduced in Chapter

3—prioritizing capability building while demonstrating value—and aligns closely with the AI Maturity Model in Chapter 4. As organizations move from exploration to industry leadership, measurement must evolve to reflect both immediate progress and long-term strategic goals. This chapter offers a framework for assessing AI's contribution holistically at every stage of the journey.

Key Areas to Measure by AI Maturity Level

Maturity Level 1: Awareness

Focus: Building foundational knowledge and organizational readiness for AI adoption

- Contribution to AI Capability and Maturity

  - Early AI projects should prioritize organizational openness, fostering expertise and establishing workflows that integrate AI effectively.

  - Outcome: Enhancing AI maturity lays the groundwork for future applications and accelerates the organization's ability to leverage AI.

  - Metrics:

    - Employee AI literacy rate

    - Organizational AI readiness assessment score

    - Number of AI awareness campaigns conducted

- Tracking Ethical and Responsible AI Use

  - Introducing ethical considerations early builds trust and reduces risks as AI initiatives expand.

- Outcome: Establishing a baseline for governance and ethical use prepares the organization for scaling AI responsibly.
- Metrics:
  - Initial bias audit results
  - Percentage of governance policies aligned with AI ethics

Maturity Level 2: Experimentation

Focus: Testing AI capabilities through pilots or proof-of-concept (PoC) projects to identify viable use cases

- Progression from PoC to Production
  - Tracking the lifecycle of PoCs helps assess technical feasibility and alignment with business goals.
  - Outcome: Successful pilots demonstrate the potential for scalable, impactful AI solutions.
  - Metrics:
    - Number of PoCs initiated
    - Percentage of PoCs successfully completed
    - Average time-to-prototype for AI solutions
- Measuring ROI (Preliminary)
  - During experimentation, metrics often focus on pre- and post-implementation comparisons for pilot projects.
  - Outcome: Provides early indications of potential financial and operational impacts.
  - Metrics:

- Reduction in operational costs for pilot workflows

- Improvements in process efficiency for targeted use cases

- Tracking Workforce Transformation (Early Stage)

  - Assess employee engagement and readiness to adopt AI during experimentation.

  - Outcome: Identifies areas in which training or process adjustments are required to support AI initiatives.

  - Metrics:

    - Number of employees involved in PoC evaluations

    - Feedback from employees on AI pilot usability

Maturity Level 3: Integration

Focus: Embedding AI solutions into operational workflows and scaling initial successes across departments

- Measuring Workforce Transformation

  - As AI is integrated, metrics focus on adoption rates and productivity improvements.

  - Outcome: Ensures AI tools are being used effectively and workflows are optimized.

  - Metrics:

    - Percentage of workflows augmented by AI

    - Employee adoption rates for AI-powered tools

    - Reduction in manual task completion times

- Alignment with Long-Term Strategic Objectives

  - AI initiatives should align with overarching business

goals like enhancing customer service or driving innovation.

- Outcome: Embedding AI in operations transforms it from an experimental capability into a strategic enabler.
- Metrics:
  - Customer satisfaction improvements (e.g., Net Promoter Score)
  - Increase in service delivery speed for AI-enhanced workflows
- Tracking Ethical and Responsible AI Use
  - Scaling AI responsibly requires continuous monitoring of ethical compliance.
  - Outcome: Maintains stakeholder trust and regulatory adherence during scaling.
  - Metrics:
    - Percentage of AI models passing fairness checks
    - Number of updates to ethical governance policies

Maturity Level 4: Optimization
Focus: Refining AI systems for maximum performance, aligning them with strategic objectives, and improving scalability

- Scalability and Resilience of AI Solutions
  - Optimization involves ensuring AI systems can handle increased complexity and adapt to evolving demands.
  - Outcome: Scalable systems enable organizations to expand their capabilities efficiently.

- Metrics:
  - Number of AI systems deployed enterprise-wide
  - Infrastructure readiness for handling increased data volumes
  - Downtime or disruption rates for AI systems
- Measuring ROI (Advanced)
  - At this stage, ROI metrics focus on long-term impacts, such as sustained cost savings and revenue growth.
  - Outcome: Demonstrates AI's financial and strategic value at scale.
  - Metrics:
    - Percentage reduction in operational costs across departments
    - Revenue growth directly attributable to AI solutions
- Measuring AI-Driven Innovation
  - Optimization drives innovation by enabling new products and services.
  - Outcome: Differentiates the organization in competitive markets.
  - Metrics:
    - Number of new products launched using AI insights
    - Revenue generated from AI-enabled innovations

Maturity Level 5: Leadership

Focus: Establishing AI as a strategic differentiator, driving innovation, market leadership, and competitive advantage

- Measuring AI-Driven Innovation
  - At this level, innovation becomes a hallmark of AI leadership.
  - Outcome: Supports market expansion and cements the organization's position as an industry leader.
  - Metrics:
    - Number of AI-related patents filed or approved
    - Expansion into new markets or customer segments through AI-enabled strategies
- Monitoring Customer Impact and Experience
  - AI leadership often emphasizes creating superior customer experiences.
  - Outcome: Builds loyalty, drives revenue, and enhances brand value.
  - Metrics:
    - Increase in customer retention rates due to AI-driven personalization
    - Reduction in complaints or negative feedback about AI-enhanced services

Intangible Benefits

While tangible metrics like ROI and cost reduction often dominate discussions about AI success, intangible benefits are equally critical for assessing AI's true value. These benefits—such as enhanced decision-making, improved employee morale, and strengthened stakeholder trust—represent the transformative potential of AI beyond immediate financial outcomes. However, their less quantifiable nature often leads them to be overlooked. This section will explore the importance of intangible benefits, how to measure them effectively, and why they matter in building long-term organizational value.

Key Intangible Benefits of AI Adoption

1. Enhanced Decision-Making from AI-Driven Insights
   - AI enables leaders to make more informed decisions by analyzing vast datasets, identifying patterns, and providing actionable recommendations.
   - Intangible Value: Improved strategic agility, faster decision-making, and the ability to preempt risks based on predictive analytics
   - Example: A healthcare organization uses AI to predict patient readmission rates, enabling administrators to allocate resources more effectively and improve patient outcomes.

2. Improved Employee Morale and Job Satisfaction
   - By automating repetitive and mundane tasks, AI allows employees to focus on more creative, strategic, and value-added activities.
   - Intangible Value: Increased job satisfaction, enhanced innovation, and reduced burnout
   - Example: A customer service team benefits from AI-powered chatbots handling routine inquiries, freeing agents to resolve complex issues that require empathy and critical thinking.

3. Strengthened Stakeholder Trust Through Ethical AI Practices

- Implementing ethical AI practices fosters trust among customers, employees, and regulators, positioning the organization as a responsible and transparent innovator.
- Intangible Value: Enhanced brand reputation, customer loyalty, and resilience to reputational risks.
- Example: A financial institution ensures fairness in its AI-driven credit approval processes, increasing customer confidence and compliance with regulatory standards.

4. Enhanced Organizational Goodwill and Market Valuation
   AI-driven improvements in trust, brand reputation, and innovative positioning strengthen the goodwill reflected on the balance sheet. These intangibles not only foster deeper customer and stakeholder loyalty, but can also elevate enterprise value in the eyes of investors. As AI initiatives build distinctive capabilities and market differentiation, they contribute to higher valuation multiples, ultimately enhancing shareholder confidence.

Methods to Measure Intangible Benefits

1. Surveys and Focus Groups
   - Why: These tools capture qualitative insights directly from stakeholders, providing a nuanced understanding of how AI impacts perceptions and satisfaction.
   - How to Apply:
     - Conduct employee surveys on morale and workload changes after AI implementation.
     - Use customer focus groups to evaluate the perceived fairness or value of AI-driven interactions.

- Example Metrics:
  - Percentage of employees reporting increased job satisfaction post-AI adoption
  - Customer sentiment scores regarding AI-powered service

2. Proxy Metrics for Organizational Impact
   - Why: Proxy metrics help quantify intangible outcomes by linking them to measurable indicators.
   - How to Apply:
     - Monitor brand sentiment through social media analysis or customer reviews.
     - Track customer loyalty through Net Promoter Scores (NPS) or repeat purchase rates.
   - Example Metrics:
     - Change in positive brand mentions after launching AI-driven personalization
     - Increase in customer retention rates following AI-driven enhancements

3. Tracking Ethical AI Perception
   - Why: Ethical AI practices are central to building trust, yet their impact is best measured through indirect indicators.
   - How to Apply:
     - Conduct stakeholder trust surveys, asking customers and employees how they perceive the fairness and transparency of AI systems.
     - Track participation in ethical AI initiatives, such as audits or training programs.
   - Example Metrics:

- Percentage of stakeholders expressing confidence in the organization's use of AI
- Number of AI models that pass ethical compliance checks during audits

Strategic Importance of Intangible Benefits

1. Long-Term Differentiation
   - Organizations that excel in leveraging intangible benefits like trust and innovation position themselves as leaders in their industries.
   - Example: A technology company with a strong reputation for ethical AI practices attracts top talent and maintains customer loyalty despite fierce competition.

2. Improved Organizational Resilience
   - Intangible benefits often serve as buffers against external shocks, such as regulatory changes or public scrutiny.
   - Example: A company with high employee morale adapts more effectively to AI-driven transformations, minimizing resistance and maintaining productivity.

3. Compounding Effects Over Time
   - Unlike immediate financial metrics, intangible benefits tend to grow and compound as they reinforce other areas of success.
   - Example: Enhanced decision-making from AI insights leads to better product launches, which in turn, improve brand reputation and market share.

Example: Measuring Intangible Benefits in Retail AI Adoption

A global retailer implemented AI-driven inventory management to optimize stock levels. While tangible benefits included cost reductions and improved efficiency, the organization also tracked intangible outcomes:

- Enhanced Decision-Making: Store managers reported higher confidence in decision-making due to real-time AI insights into stock levels.

- Improved Employee Morale: Employees expressed satisfaction with reduced manual inventory checks, allowing them to focus on customer engagement.

- Strengthened Trust: Customers perceived the retailer as innovative, citing fewer instances of out-of-stock items as a positive shopping experience.

Intangible benefits are a cornerstone of AI's transformative potential, shaping how organizations grow, innovate, and build trust. By using tools like surveys, focus groups, and proxy metrics, organizations can capture these outcomes and integrate them into their measurement frameworks. While less immediately visible than financial gains, intangible benefits often hold the key to sustaining AI-driven success over the long term.

Industry-Specific Metric Examples

AI adoption varies widely across industries, with each sector presenting unique opportunities and challenges. Tailoring metrics to specific industries not only makes measurement frameworks more actionable, but also helps organizations contextualize success based on their priorities and maturity levels. We will explore the following key metrics for healthcare, finance, and retail, illustrating how these metrics evolve with organizational maturity.

Healthcare

AI is revolutionizing healthcare by improving diagnostics, personalizing treatments, and streamlining administrative processes. Metrics in this sector focus on patient outcomes, operational efficiency, and compliance.

1. AI Accuracy in Diagnostic Predictions
   - Why It Matters: Measures the effectiveness of AI in diagnosing conditions such as cancer or heart disease compared to traditional methods.
   - Metric Example: Percentage of correct diagnoses made by AI systems compared to human doctors.
   - Maturity Impact:
     - At lower maturity levels, focus on proof-of-concept validation for AI diagnostic tools.
     - At higher levels, track system performance across large-scale deployments.

2. Patient Satisfaction with AI-Driven Interactions
   - Why It Matters: Evaluates how well AI enhances patient experiences, such as through AI chatbots or virtual health assistants.
   - Metric Example: Improvement in patient satisfaction scores after introducing AI-powered appointment scheduling or symptom-checking tools

- Maturity Impact:
  - In early stages, track initial adoption rates and patient feedback.
  - In advanced stages, measure long-term impacts on patient retention and care quality.

3. Operational Efficiency in Administrative Tasks
   - Why It Matters: Automates time-consuming processes like billing, insurance claims, and patient record management.
   - Metric Example: Time saved in processing claims using AI-driven automation.
   - Maturity Impact:
     - In early stages, focus on time-to-implementation of AI tools.
     - At higher maturity levels, track the percentage reduction in administrative costs.

Finance

In finance, AI is widely used for fraud detection, credit risk analysis, and personalized customer service. Metrics focus on trust, security, and efficiency.

1. Reduction in Fraud Rates Due to AI
   - Why It Matters: Measures the success of AI in identifying and preventing fraudulent transactions.
   - Metric Example: Percentage decrease in undetected fraudulent activities after implementing AI-powered fraud detection
   - Maturity Impact:
     - Early stages emphasize PoC success in identifying

fraud patterns.
- Later stages focus on scalability and reducing false positives.

2. Time Saved in Loan Processing via Automation
   - Why It Matters: Tracks how AI accelerates traditionally manual processes like credit scoring and loan approvals.
   - Metric Example: Average reduction in loan processing times after implementing AI-driven credit scoring models.
   - Maturity Impact:
     - Initial focus on pilot testing with specific loan categories.
     - Advanced metrics include the percentage of loans processed without human intervention.

3. Customer Retention Through Personalized Services
   - Why It Matters: AI-powered personalization strengthens customer relationships by tailoring financial products and services.
   - Metric Example: Increase in customer retention rates due to AI-based recommendations for investment portfolios or credit products.
   - Maturity Impact:
     - Early stages track pilot adoption rates.
     - Higher levels focus on AI's impact on overall revenue and market share.

Retail

Retail organizations use AI to personalize customer experiences, optimize inventory, and improve supply chain efficiency. Metrics center on revenue growth and operational performance.

1. Revenue Increase from Personalized AI Recommendations
   - Why It Matters: Tracks AI's contribution to sales by offering tailored product recommendations.
   - Metric Example: Percentage of revenue attributed to AI-powered upselling and cross-selling.
   - Maturity Impact:
     - Initial focus on customer engagement with AI-driven recommendations.
     - Advanced tracking involves the correlation between personalization and customer lifetime value (CLV).

2. Inventory Optimization Rates
   - Why It Matters: Measures AI's ability to forecast demand and reduce excess inventory.
   - Metric Example: Reduction in stockouts and overstock situations through AI-driven demand forecasting.
   - Maturity Impact:
     - At lower maturity levels, measure AI's accuracy in predicting demand.
     - At higher levels, evaluate cost savings and customer satisfaction improvements.

3. Customer Satisfaction with AI-Powered Services
   - Why It Matters: Assesses how AI enhances customer experiences, such as through virtual shopping assistants or chatbots.

- Metric Example: Improvement in Net Promoter Scores (NPS) or CSAT after deploying AI-powered customer support.
- Maturity Impact:
  - Early stages measure adoption rates of AI-powered tools.
  - Advanced levels focus on long-term loyalty and repeat purchase rates.

The Role of Maturity in Industry-Specific Metrics

Organizational maturity significantly influences which metrics are most relevant and how they should be applied.

- Awareness and Experimentation Stages: Metrics focus on feasibility, adoption rates, and initial feedback.

- Integration and Optimization Stages: Metrics shift toward efficiency gains, scalability, and stakeholder satisfaction.

- Leadership Stage: Emphasis is placed on innovation, competitive differentiation, and long-term strategic outcomes.

By tailoring metrics to their respective industry and maturity level, organizations can derive actionable insights and maximize the value of their AI initiatives.

Challenges in Defining and Tracking Metrics

Measuring the success of AI initiatives is an inherently complex task. Unlike traditional technologies with well-defined outcomes, AI's impact is multifaceted, dynamic, and often interdependent with other organizational efforts. This makes defining and tracking meaningful metrics challenging, particularly as organizations move through different stages of AI maturity. In the following segment, we will explore key challenges in establishing effective metrics and provide actionable best practices to navigate these complexities.

Key Challenges in Defining and Tracking Metrics

1. Difficulty Isolating AI's Impact from Other Variables
   - AI rarely operates in isolation; its outcomes are often intertwined with other factors such as process improvements, market dynamics, or human decision-making.
   - For example, a rise in customer satisfaction may result from a combination of AI-powered personalization, improved customer service training, and broader product enhancements.

2. Metrics That Evolve Over Time
   - As organizations progress in AI maturity, their priorities and goals shift, requiring corresponding changes in how success is measured.
   - In the Experimentation stage, metrics may focus on technical feasibility (e.g., PoC success rates), whereas in the Leadership stage, metrics emphasize innovation (e.g., patents filed or new market expansion).

3. Balancing Long-Term Outcomes with Short-Term Gains
   - Organizations often face pressure to demonstrate quick wins to justify AI investments, which can overshadow the importance of long-term value creation.
   - For instance, prioritizing cost reduction in the short term may undermine efforts to build capabilities necessary for future scalability.

4. Subjectivity in Measuring Intangible Benefits
   - Intangible outcomes, such as enhanced decision-making, workforce empowerment, or brand reputation, are difficult to quantify and may be overlooked in favor of easily measurable metrics.

5. Data Quality and Governance Issues
   - Effective metrics depend on high-quality data, which can be compromised by inconsistencies, biases, or gaps in data governance.
   - Poor data management can lead to misleading metrics, eroding trust in AI systems and decision-making processes.

Best Practices for Overcoming Challenges

1. Use Baseline Comparisons to Isolate AI's Impact
   - Establish clear baseline measurements for key metrics before implementing AI solutions.
   - Example: Compare customer satisfaction scores pre- and post-AI-powered chatbot implementation to isolate the chatbot's specific impact.

- Ensure baseline data is robust and accounts for external variables (e.g., market seasonality or concurrent organizational changes).

2. Build Flexible, Adaptive Measurement Frameworks
   - Design metrics frameworks that evolve with the organization's AI maturity and strategic goals.
   - Example: In early stages, focus on metrics like employee AI literacy rates or PoC success. In later stages, shift toward operational efficiency, customer impact, and innovation metrics.
   - Regularly review and adjust metrics to ensure relevance and alignment with organizational priorities.

3. Incorporate Both Quantitative and Qualitative Measures
   - Balance tangible metrics (e.g., revenue growth, cost savings) with qualitative insights (e.g., employee surveys, customer feedback).
   - Example: Combine employee engagement surveys with productivity metrics to assess workforce transformation comprehensively.
   - Use techniques like focus groups or interviews to capture nuanced, intangible benefits such as trust in AI systems or ethical perceptions.

4. Address Data Quality Proactively
   - Implement robust data governance practices to ensure data accuracy, consistency, and representativeness.
   - Example: Conduct regular audits of AI training datasets to identify and address biases or gaps.
   - Use automated tools for real-time monitoring of

data pipelines, reducing the risk of errors in metrics calculations.

5. Communicate Metrics Contextually to Stakeholders
   • Present metrics in a way that aligns with the audience's priorities and understanding.
   • Example: For executives, highlight high-level outcomes like ROI or market share growth. For operational teams, focus on granular metrics like process efficiency or adoption rates.
   • Provide context around metrics, explaining how they tie to broader goals and addressing potential limitations or nuances.

Example: Navigating Metric Complexities in Retail AI Adoption

A retail organization implemented AI to enhance customer experiences through personalized recommendations. Key challenges arose in isolating AI's impact, as other improvements (e.g., faster delivery times, product enhancements) also influenced customer satisfaction.

Approach

1. The organization established pre-AI baselines for customer satisfaction and engagement scores.

2. They implemented qualitative feedback mechanisms, like surveys, to understand customer perceptions of the AI recommendations specifically.

3. Metrics frameworks were adjusted as the initiative matured, shifting focus from AI adoption rates to revenue increases attributable to AI-driven personalization.

Outcome

By addressing these challenges with adaptive and context-aware metrics, the organization demonstrated AI's unique contributions to both short-term gains and long-term strategic objectives.

Defining and tracking AI metrics requires navigating complexities that stem from AI's dynamic and multifaceted nature. By recognizing common challenges—such as isolating AI's impact, evolving metrics, and balancing short- and long-term goals—organizations can proactively address potential pitfalls. Employing best practices like baseline comparisons, flexible frameworks, and a balance of quantitative and qualitative measures ensures metrics remain relevant, actionable, and aligned with strategic priorities.

Applying These Metrics

While we have presented these metrics aligned with each stage of maturity for simplicity, real life is rarely this straightforward. Different parts of an organization often reach varying levels of AI maturity simultaneously, reflecting unique challenges, priorities, and resource availability. Additionally, some metrics naturally span multiple maturity levels—such as customer satisfaction or ethical compliance—which remain relevant throughout the AI journey.

Successful application of these frameworks relies on thoughtful customization and the use of sound managerial judgment. Organizations should adapt these metrics to reflect their specific context, ensuring alignment with their strategic goals and the current state of their AI initiatives. By tailoring these guidelines, businesses can derive meaningful insights and drive sustainable value from their AI investments.

Chapter Summary

This chapter underscored the importance of aligning AI metrics with both strategic goals and organizational maturity, ensuring relevance across evolving priorities. It highlighted the need for adaptability in measurement frameworks to address real-world complexities and interdependencies. By balancing tangible outcomes with intangible benefits, organizations can uncover AI's transformative potential. The approach empowers decision-makers to leverage metrics as strategic tools for sustained innovation and competitive advantage.

Actionable Takeaway

Develop a metrics framework tailored to your organization's AI initiatives.

Next Chapter

Metrics tell you where you have been, but staying ahead in AI means knowing where you are going next—and how to keep moving. The final chapter will explore what it takes to lead when the pace of change never slows.

Chapter 9

Your Journey Forward

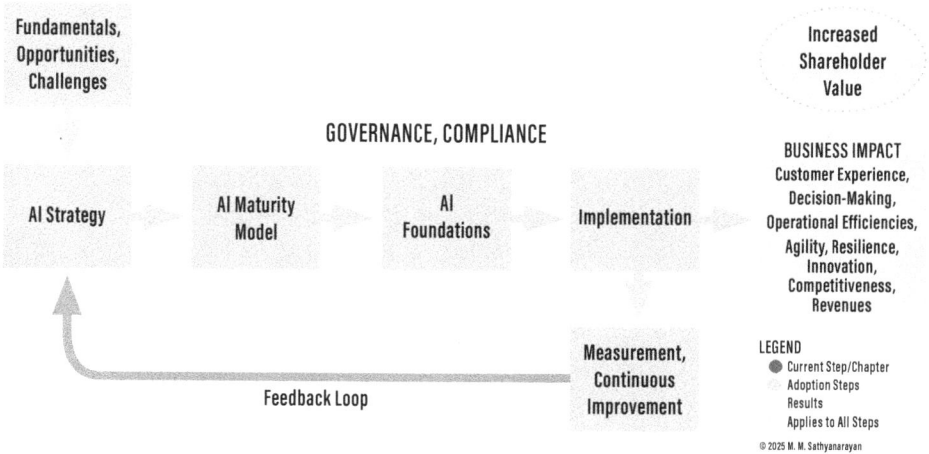

FIGURE 9.1 AI ADOPTION ROADMAP

The journey to AI maturity doesn't end with implementation; it requires a commitment to continuous growth and adaptability (see Figure 9.1). This chapter explores how to sustain momentum by fostering a culture of innovation, leveraging networks, institutionalizing continuous learning, and integrating lessons learned into future strategies. By embracing these practices, your organization can ensure long-term success in an ever-evolving AI landscape.

Stay Abreast of AI Developments

The AI landscape evolves rapidly, with advancements in technologies, frameworks, and regulations. Organizations must remain proactive in tracking these changes to ensure competitiveness and compliance. This isn't just about staying informed—it's about building an agile organization that can adapt to change and seize emerging opportunities. It also means cultivating a disciplined "AI first" mindset at every level, consistently asking: *What's out there today? How can we use it? Should we buy, build, or partner?* Embedding this habit ensures AI becomes a default lens for solving business challenges, not an afterthought. These practices build on the OODA loop (Observe, Orient, Decide, Act) approach from Chapter 3—turning environmental scanning into a disciplined, repeatable cycle that keeps your organization adaptive and ahead of the curve.

How to Stay Updated

- Institutionalize Learning Mechanisms
  Establish regular forums or "AI knowledge-sharing sessions" within your organization in which teams can present recent developments in AI technology or regulations.

- Encourage Collaboration Across Teams
  Create cross-departmental committees that track AI trends and analyze their relevance to the organization. For example, a "Future of AI" task force can explore new tools or frameworks and provide recommendations.

- Leverage External Resources
  Participate in AI consortiums or regulatory working groups, as they provide early access to industry-wide shifts and best practices.

Cultivate Innovation Culture

An innovation culture ensures that AI efforts are scalable and resilient to future demands. But fostering innovation isn't a one-time effort; it requires creating an environment in which experimentation, collaboration, and learning thrive continuously.

Key Strategies

- Reward Experimentation
  Celebrate successes and learn from failures. For instance, set up an "AI Lab" in which teams can prototype new ideas without fear of failure. This approach aligns with the iterative principles discussed in Chapter 3, such as prioritizing capability building.

- Promote Leadership Engagement
  Leaders should champion AI innovation by providing clear direction and support. They can do this by allocating budgets for experimentation and publicly recognizing teams that push boundaries.

- Embed Lessons Learned into Future Efforts
  After completing AI initiatives, conduct retrospectives to identify what worked, what didn't, and why. These lessons can inform the next round of projects, ensuring continuous improvement.

Connecting to Earlier Chapters

This strategy builds on the guiding principles from Chapter 3, particularly fostering a culture of continuous learning and iteration. The AI Maturity Model in Chapter 4 also emphasizes the importance of scaling innovation as organizations advance.

Leverage Community and Network

AI adoption is a team sport. Collaborating with external and internal stakeholders accelerates innovation and ensures alignment with broader industry practices. By engaging with the broader AI ecosystem, organizations can gain valuable insights and avoid reinventing the wheel.

How to Leverage Networks

- Internal Knowledge Networks
  Create internal AI working groups that include representatives from multiple departments. These groups can share insights, align on best practices, and ensure that AI efforts remain consistent across the organization.

- External Collaboration
  - Partner with universities to conduct AI research relevant to your industry.
  - Collaborate with technology providers to co-develop tailored AI solutions.

- Participate in Global AI Forums
  Joining forums like the Partnership on AI or regional consortia allows organizations to shape industry practices and gain early access to advances in technology.

Connecting to Earlier Chapters
Building external and internal networks ties directly to the collaborative strategies highlighted in Chapter 3. These networks also accelerate maturity progression, as discussed in Chapter 4, by enabling knowledge sharing and resource pooling.

Institutionalizing Continuous Learning

To remain competitive in the dynamic AI landscape, organizations must embed continuous learning into their core operations. This isn't just about individual skill-building; it's about creating systems that promote organization-wide growth and adaptability.

Strategies to Institutionalize Learning

- Establish a Learning Ecosystem
  Develop AI-specific training programs for employees at all levels. For example, offer beginner-friendly courses for nontechnical teams and advanced certifications for technical staff.

- Rotate Talent Across AI Projects
  Rotating employees through different AI initiatives helps build institutional knowledge and cross-functional expertise.

- Appoint AI Champions
  Identify key individuals in each department to serve as AI ambassadors. These champions can drive adoption, track trends, and promote best practices.

Connecting to Earlier Chapters

This focus on learning echoes the strategic emphasis on building AI capability in Chapter 3 and supports the ongoing progression outlined in Chapter 4. By institutionalizing learning, organizations ensure that they not only adopt AI, but also thrive in its evolution.

Chapter Summary

Sustaining AI maturity hinges on embedding continuous learning and adaptability into your organization's culture. By fostering innovation and leveraging internal and external networks, you position your organization to thrive amid AI's evolving opportunities and challenges. These practices ensure lasting impact and resilience in an ever-changing landscape.

Actionable Takeaway

Create a sustainable learning framework within your organization to stay relevant in the rapidly changing AI ecosystem.

Bibliography

Accenture. (July 2020). AI: An Engine for Growth. Retrieved from https://www.accenture.com/au-en/insights/health/artificial-intelligence-healthcare

Accenture. (2024). Technology Vision 2024: Meet Me in the Metaverse—Shaping the Future of Work with AI Agents. Retrieved from https://www.accenture.com/us-en/insights/technology/technology-trends-2024

Bhaskar Ghosh, R. P. (2022). How to Pick the Right Automation Project. Retrieved from https://hbr.org/2022/02/how-to-pick-the-right-automation-project

Daugherty, J. W. (2018). Collaborative Intelligence: Humans and AI Are Joining Forces. *Harvard Business Review*. Retrieved from Harvard Business Review: https://hbr.org/2018/07/collaborative-intelligence-humans-and-ai-are-joining-forces

Deloitte Consulting. (2022). AI Is an Increasingly Critical Tool, Health Leaders Say. Retrieved from https://www2.deloitte.com/us/en/blog/health-care-blog/2022/ai-is-an-increasingly-critical-tool-health-leaders-say.html

DeNittis, N. (2022). Early AI Adoption: How to Avoid Classic Mistakes—With David Carmona of Microsoft. (Emerj). Retrieved from https://emerj.com/early-ai-adoption-avoiding-classic-mistakes/

Emerj. (2020). AI in Supply Chain and Logistics: DHL Case Study. Sourced from Emerj. www. Emerj.com. Retrieved from https://emerj.com/ artificial-intelligence-at-dhl-two-applications/

EY. (n.d.). How Generative AI Can Drive Value in Supply Chain. Retrieved from https://www.ey.com/en_us/coo/ how-generative-ai-in-supply-chain-can-drive-value

Iavor Bojinov, H. (2023). Keep Your AI Projects on Track. *Harvard Business Review.* Retrieved from Harvard Business Review: https://hbr.org/2023/11/ keep-your-ai-projects-on-track

Jarrell, M. (2018). How Amazon Uses AI in eCommerce—Two Use-Cases. Emerj Artificial Intelligence Research. Retrieved from https://emerj.com/ai-sector-overviews/ how-amazon-uses-ai-in-ecommerce-two-use-cases/

Juniper Research. (2023). Voice Commerce: The Rise of Voice-Activated Shopping in eCommerce. TechBullion, 2023. Retrieved from https://techbullion.com/ voice-commerce-rise-of-voice-activated-shopping/

Microsoft. (n.d.). CarMax Puts Customers First with Car Research Tools Powered by Azure OpenAI Service. Retrieved from https://www.microsoft.com/en/customers/sto- ry/1501304071775762777-carmax-retailer-azure-ope- nai-service

Microsoft. (n.d.). Copilot. Introducing Microsoft 365 Copilot. Microsoft 365 Blog.

Nemteanu, I. (2018). Data Science Processes and Standards, Nemsee, LLC.